Military Strategies of the New European Allies

This book analyses how and to what extent ex-communist states have adjusted their defence strategies since joining the European Union (EU) and NATO and how differences and similarities between their strategies can be explained.

Between 1999 and 2013, four phases of enlargement took place when the EU and NATO allowed 11 new former communist states to enter both organisations. These states share some common attributes and experiences related to strategic culture and common experiences during the Cold War era that can potentially explain similarities in behaviour and preferences among them. However, the strategic adjustments among these states are far from uniform. In an effort to explain these differences, the book introduces three intervening variables: (1) differences in relative power and position in the international system, (2) national geographical characteristics, and (3) historical experiences related to formative periods of state-building processes as well as wars and armed conflicts. Empirically, the book strives to present and analyse the defence strategies of each of the new allies by conducting a structured focused comparison of official strategic documents from the twenty-first century for each of the 11 cases. Theoretically and methodologically, it introduces an analytical framework enabling us to explain both similarities and differences in the formulation of the strategies of the 11 states and to shed light on their external and internal efforts to promote their strategic interest by operationalising the dependent variable – defence strategy. The analytical framework combines elements of structural realism with classical realism, as well as constructivist research on unit-level characteristics linked to the relative power and perceptions of strategic exposure.

This book will be of great interest to students of Strategic Studies, European Union policy, NATO and international relations in general.

Håkan Edström is an associate professor in political science and a senior lecturer in war studies at the Swedish Defence University, Stockholm.

Jacob Westberg is an associate professor in war studies and a senior lecturer in security policy and strategy at the Swedish Defence University, Stockholm.

Cass Military Studies

Counterinsurgency Warfare and Brutalisation
The Second Russian-Chechen War
Roberto Colombo and Emil Aslan Souleimanov

Managing Security
Concepts and Challenges
Edited by Laura R. Cleary and Roger Darby

Understanding the Impact of Social Research on the Military
Reflections and Critiques
Edited by Eyal Ben-Ari, Helena Carreiras, and Celso Castro

Civil-Military Cooperation in International Interventions
The Role of Soldiers
Agata Mazurkiewicz

Contemporary Military Reserves
Between the Civilian and Military Worlds
Edited by Eyal Ben-Ari and Vincent Connelly

Military Strategies of the New European Allies
A Comparative Study
Håkan Edström and Jacob Westberg

Proxy War in Yemen
Bernd Kaussler and Keith A. Grant

For more information about this series, please visit: www.routledge.com/Cass-Military-Studies/book-series/CMS

Military Strategies of the New European Allies
A Comparative Study

Håkan Edström and Jacob Westberg

LONDON AND NEW YORK

First published 2023
by Routledge
4 Park Square, Milton Park, Abingdon, Oxon OX14 4RN

and by Routledge
605 Third Avenue, New York, NY 10158

Routledge is an imprint of the Taylor & Francis Group, an Informa business

© 2023 Håkan Edström and Jacob Westberg

The right of Håkan Edström and Jacob Westberg to be identified as authors of this work has been asserted in accordance with sections 77 and 78 of the Copyright, Designs and Patents Act 1988.

All rights reserved. No part of this book may be reprinted or reproduced or utilised in any form or by any electronic, mechanical, or other means, now known or hereafter invented, including photocopying and recording, or in any information storage or retrieval system, without permission in writing from the publishers.

Trademark notice: Product or corporate names may be trademarks or registered trademarks, and are used only for identification and explanation without intent to infringe.

British Library Cataloguing-in-Publication Data
A catalogue record for this book is available from the British Library

ISBN: 978-1-032-28693-8 (hbk)
ISBN: 978-1-032-28695-2 (pbk)
ISBN: 978-1-003-29805-2 (ebk)

DOI: 10.4324/9781003298052

Typeset in Times New Roman
by Apex CoVantage, LLC

Contents

List of Tables vii
List of Abbreviations viii

PART I
**Theoretical and methodological considerations:
framing the research design** 1

1 Strategic adjustment in Central and Eastern Europe 3
2 Analysing and explaining strategic adjustment and diversity 14
3 Operationalising the dependent variable: defence strategy 39

PART II
The empirical exploration 49

4 The strategy of Bulgaria 51
5 The strategy of Croatia 61
6 The strategy of the Czech Republic (Czechia) 71
7 The strategy of Estonia 82
8 The strategy of Hungary 93
9 The strategy of Latvia 104
10 The strategy of Lithuania 115
11 The strategy of Poland 126
12 The strategy of Romania 138

13 The strategy of Slovakia 150

14 The strategy of Slovenia 161

PART III
Explaining the findings 173

15 The aggregated result of the empirical exploration 175

16 Explaining the diversity of strategic responses 182

17 Conclusions: strategic responses to membership demands and changes in the external security environment 206

Index 216

Tables

1.1	New European allies.	4
2.1	Average GDP and GDP per capita 2010–2019.	26
2.2	Average military expenditures 2010–2019.	26
2.3	Foreign embassies in the 11 new European allies.	27
2.4	Potential aggregated perception of strategic exposure.	33
3.1	Alignment strategies.	41
3.2	Elements of military strategy.	46
4.1	Main military resources of Bulgaria.	56
4.2	Bulgarian strategy.	58
5.1	Main military resources of Croatia.	66
5.2	Croatian strategy.	68
6.1	Main military resources of Czechia.	76
6.2	Czech strategy.	78
7.1	Main military resources of Estonia.	86
7.2	Estonian strategy.	89
8.1	Main military resources of Hungary.	98
8.2	Hungarian strategy.	101
9.1	Main military resources of Latvia.	109
9.2	Latvian strategy.	111
10.1	Main military resources of Lithuania.	120
10.2	Lithuanian strategy.	122
11.1	Main military resources of Poland.	132
11.2	Polish strategy.	134
12.1	Main military resources of Romania.	144
12.2	Romanian strategy.	146
13.1	Main military resources of Slovakia.	155
13.2	Slovakian strategy.	157
14.1	Main military resources of Slovenia.	166
14.2	Slovenian strategy.	168
15.1	Strategies of the 11 new European allies.	176
16.1	WWII experiences of armed aggression.	196
16.2	Intervening variables and the outcomes of the defence strategy.	201
17.1	Number of MBTs and combat aircraft of some of the new allies.	210

Abbreviations

ACV	armoured combat vehicles
AFCR	Armed Forces of the Czech Republic
ALD	Alliance of Liberals and Democrats (political party of Romania)
APC	armoured personnel carrier
AWS	Solidarity Electoral Action (political party of Poland)
BLACKSEAFOR	Black Sea Naval Cooperation Group
BSP	Bulgarian Socialist Party (political party of Bulgaria)
CFSP	Common Foreign and Security Policy (of the EU)
CIS	Commonwealth of Independent States
CM	Council of Ministers
COPD	Comprehensive Operations Planning Directive (of NATO)
CRBN	chemical, biological, radiological and nuclear
CSDP	Common Security and Defence Policy (of the EU)
ČSSD	Social Democratic Party (political party of the Czech Republic)
DWP	defence white paper
EAEC	European Atomic Energy Community (Euratom)
EC	European Communities
ECSC	European Coal and Steel Community
EEC	European Economic Community
EKE	Centre Party (political party of Estonia)
ERE	Reform Party (political party of Estonia)
ESDP	European Security and Defence Policy (of the EU)
EU	European Union
EUBG	EU Battle Group
EUFOR	EU Force
EUMM	EU Monitoring Mission
EUPOL	EU Police Mission
EUTM	EU Training Mission
FRG	Federal Republic of Germany (West Germany)
G7	Group of Seven
G20	Group of Twenty

GDP	gross domestic product
GERB	Citizens for European Development of Bulgaria (political party of Bulgaria)
GS	General Staff
HDZ	Democratic Union (political party of Croatia)
HNS	host nation support
IFV	infantry fighting vehicle
IISS	International Institute for Strategic Studies
IMF	International Monetary Fund
IR	International Relations
IS	Islamic State
ISAF	International Security Assistance Force (of NATO)
JL	New Era Party (political party of Latvia)
JV	New Unity (political party of Latvia)
KFOR	Kosovo Force (of NATO)
LC	Latvian Way (political party of Latvia)
LDS	Liberal Democracy of Slovenia (political party of Slovenia)
LMŠ	List of Marjan Šarec (political party of Slovenia)
LP	Liepāja Party (political party of Latvia)
LSDP	Lithuanian Social Democratic Party (political party of Lithuania)
LVZS	Farmers and Greens Union (political party of Lithuania)
LZP	Green Party (political party of Latvia)
MA	military assistance
MAP	Membership Action Plan (of NATO)
MBT	main battle tank
MC	military cooperation
MCMV	mine countermeasures vessel
MFEA	Ministry of Foreign and European Affairs
MINURCAT	UN Mission in the Central African Republic and Chad
MLF	Multinational Land Force (of NATO)
MNC-NE	Multinational Corps Northeast (of NATO)
MoD	Ministry of Defence
MoDA	Ministry of Digital Affairs
MoFA	Ministry of Foreign Affairs
MPF-SEE	Multinational Peace Force–South-Eastern Europe
MSZP	Socialist Party (political party of Hungary)
NA	National Assembly
NATO	North Atlantic Treaty Organization
NBC	nuclear, biological and chemical (weapons)
NC	National Council
NCOC	National Cyber Operations Centre
NCSS	national cyber security strategy
NDS	national defence strategy
NM	National Movement (political party of Bulgaria)

NMSP	National Movement for Stability and Progress (political party of Bulgaria)
NRDC-It	NATO Rapid Deployable Corps in Italy
NRF	NATO Response Force
NSA	National Security Authority
NSB	National Security Bureau
NSi	New Slovenia–Christian Democrats (political party of Slovenia)
NSS	national security strategy
ODS	Civic Democratic Party (political party of the Czech Republic)
OECD	Organization for Economic Co-operation and Development
OL'aNO	Ordinary People and Independent Personalities (political party of Slovakia)
OSCE	Organization for Security and Co-operation in Europe
OW	offensive warfare
PARP	Planning and Review Process (of NATO)
PDL	Democratic Liberal Party (Political Party of Romania)
PESCO	Permanent Structured Cooperation (of the EU)
PfP	Partnership for Peace (of NATO)
PiS	Law and Justice (Political Party of Poland)
PNL	National Liberal Party (political party of Romania)
PO	Civic Platform (political party of Poland)
PRT	Provincial Reconstruction Team (of NATO)
PSD	Social Democratic Party (political party of Romania)
PSO	peace support operations
RP	Res Publica Party (political party of Estonia)
RSCT	Regional Security Complex Theory
SAR	search and rescue
SD	Social Democrats (political party of Slovenia)
SDKÚ-DS	Democratic and Christian Union–Democratic Party (political party of Slovakia)
SDP	Social Democratic Party (political party of Croatia)
SDS	Slovenian Democratic Party (political party of Slovenia)
SEEBRIG	South Eastern Europe Brigade
SFC	structured focused comparison
SFOR	Stabilisation Force (of NATO)
SHIRBRIG	Multinational Stand-by High Readiness Brigade
SIPRI	Stockholm International Peace Research Institute
SLD	Democratic Left Alliance (political party of Poland)
SMC	Modern Centre Party (political party of Slovenia)
SMER-SD	Direction–Social Democracy (political party of Slovakia)
SSD	State Security Department
TP	People's Party (political party of Latvia)

TS-LKD	Homeland Union–Lithuanian Christian Democrats (political party of Lithuania)
UDF	Union of Democratic Forces (political party of Bulgaria)
UK	United Kingdom
UN	United Nations
UNDOF	UN Disengagement Observer Force
UNFICYP	UN Peacekeeping Force in Cyprus
UNIFIL	UN Interim Force in Lebanon
UNPR	National Union for the Progress of Romania (political party of Romania)
UNSC	UN Security Council
US	United States of America
USSR	Soviet Union
V	Unity (political party of Latvia)
WMD	weapons of mass destruction
WP	Warsaw Pact
WTO	World Trade Organization
WWI	First World War
WWII	Second World War

Part I
Theoretical and methodological considerations

Framing the research design

1 Strategic adjustment in Central and Eastern Europe

> From Stettin in the Baltic to Trieste in the Adriatic, an iron curtain has descended across the Continent. Behind that line lie all the capitals of the ancient states of Central and Eastern Europe. Warsaw, Berlin, Prague, Vienna, Budapest, Belgrade, Bucharest and Sofia, all these famous cities and the populations around them lie in what I must call the Soviet sphere, and all are subject in one form or another, not only to Soviet influence but to a very high and, in many cases, increasing measure of control from Moscow.
>
> Winston Churchill (2003:420)

Almost three years after Sir Winston Churchill's 'Iron Curtain Speech' in Fulton, Missouri, on 5 March 1946, 12 states signed the Washington Treaty and thus became the founding members of the North Atlantic Treaty Organization (NATO). During the Cold War, four other states joined the Alliance. In 1951, six West European states signed the Treaty of Paris, hence establishing the European Coal and Steel Community (ECSC). Six years later, the very same six countries signed the Treaty of Rome and hereby established both the European Economic Community (EEC), and the European Atomic Energy Community (EAEC, or Euratom). In 1965, the six signed the Treaty of Brussels, also known as the Merger Treaty, unifying the institutions of the ECSC, the EEC and the EAEC as the European Communities (EC).[1] During the Cold War, six other states joined the EC.[2] Consequently, when, the dissolution of the Soviet Union (USSR) eventually ended the Cold War in December 1991, NATO had 16 members while the EC had 12.[3] Notably, 11 of these states were members of both organisations.

Between 1999 and 2013, four phases of enlargement took place when the European Union (EU) and NATO allowed 11 new members to enter both organisations. In this study, we (1) map and analyse the defence strategies the 11 new allies have formulated when joining the organisations established on the Western side of this curtain and (2) develop and apply an analytical framework to analyse and explain differences and similarities among the strategies of these new allies. See Table 1.1.

In joining the EU and NATO, the 11 former communist states were exposed to common systematic pressures from the regional and global levels. During the

4 *Theoretical and methodological considerations*

Table 1.1 New European allies.[4]

	Phase of enlargement	NATO member	EU member
Czechia	First	1999	2004
Hungary	First	1999	2004
Poland	First	1999	2004
Estonia	Second	2004	2004
Latvia	Second	2004	2004
Lithuania	Second	2004	2004
Slovakia	Second	2004	2004
Slovenia	Second	2004	2004
Bulgaria	Third	2004	2007
Romania	Third	2004	2007
Croatia	Fourth	2009	2013

first decade of the twenty-first century, both the EU and NATO exercised institutional pressures on their members as well as their partners to contribute to multilateral military operations outside Europe and to develop capabilities related to expeditionary warfare including peace support operations (PSOs). This created a common challenge to the new allies to develop military capabilities that are very different from the capabilities developed and used during the Cold War era (Edmunds and Malesic 2005). At the time, the region was commonly believed to be characterized by a new 'post-national security paradigm' with an absence of existential threats to states' territorial sovereignty from other states (Matlary 2009).

Russia's military intervention in Georgia in 2008, its annexation of the Crimea in 2014 and not least its full-scale war against Ukraine in 2022, challenge these ideas of a stable peaceful European security order. NATO responded to Russia's military aggression by reactivating its efforts for common defence planning and collective defence. Within the EU, questions related to collective defence and active support to Ukraine have gained increased importance. Moreover, the new allies are particularly exposed in relation to the two proposals on security agreements with the US and NATO presented by Russia in December 2021. In these proposed agreements, it is suggested that both Russia and the US shall refrain from deploying armed forces and armaments in areas where such deployment could be perceived as a threat to the other states' national security. Moreover, states that have joined NATO before 1997 should not deploy military forces and weaponry on the territory of any other states in Europe in addition to the forces stationed on that territory as of May 1997 (Russian MoFA 2021a, 2021b). Russia's proposals indicate a renewed ambition to recreate the sphere of influence it had during the Cold War and reverse the outcome of NATOs enlargements since 1997. In addition, Russia's full-scale invasion of Ukraine in February 2022 displayed a preparedness to use military force on a scale unseen in Europe since the end of the Second World War (WWII). How have the 11 new allies responded to the systematic pressures emanating from their membership in the two alliances and the changes in their external security environment during the two first decades of the twenty-first century?

The 11 states share some common attributes and experiences related to strategic culture.[5] According to Ken Booth, strategic culture can be defined as 'a nation's traditions, values, attitudes, patterns of behaviour, habits, customs, achievement and particular ways of adapting to the environment and solving problems with respect to the use of force' (Booth 1990:121; see also Gray 1999). Another scholar, Alistair Iain Johnston, has argued that strategic culture establishes pervasive and long-lasting grand strategic preferences by formulating concepts of the role and efficacy of military force in interstate political affairs (Johnston 1995a). The 11 European states included in this study share some common characteristics and experiences that may make us expect similarities in their strategic responses to the systematic pressures emanating from their common external security environment. First, prior to the Treaty of Berlin in 1878, they were all part of one or several of the four imperial states then ruling Central and Eastern Europe, that is the Austrian, the German, the Ottoman and the Russian Empires respectively. Second, they were all theatres of brutal war fighting during WWII. Third, they are all 'post-communist states', and many of them share common historical experiences from the Cold War era as members of the Warsaw Pact (WP). However, the strategic responses from the 11 new allies have been far from uniform. In the next chapter, we introduce three intervening variables to address these differences: (1) differences in relative power and position in the system, (2) national geographical characteristics and (3) unique historical experiences related to experiences during armed conflicts and formative periods of state-building processes.

The defence strategies of the 11 new allies have not yet been comprehensively addressed. Aspects of the security and defence policy of the individual countries are most often to be found in academic journals with a specific focus on Central and Eastern Europe, such as *East European Politics and Societies*, *Journal on Baltic Security*, *Journal of Slavic Military Studies*, and *Polish Quarterly of International Affairs*. Occasionally, articles exploring individual Central and East European countries can be found in 'Western' journals such as *Armed Forces & Society*, *Comparative Strategy* and *Contemporary Security Policy*.

From time to time, studies with a comparative approach, covering more than just one country, are published. However, most often, these comparisons cover, as does Miroslav Hadžić and his colleagues' (2010), only a group of the Central and East European countries such as the Baltic States, the Visegrád Group or the Balkans rather than all of the 11 new allies explored in this project. Monographs going in-depth when analysing the defence policy of individual Central and East European countries are rare. In addition, these projects, such as Justyna Zajac's (2016) as well as the other works previously touched upon, tend to focus on security policy rather than on strategy. Such also is the case with most edited volumes. Anna Péczeli (2019) focused, for example, on the relations of the Central European countries with the United States of America (US). Moreover, in her volume, the countries are addressed individually. Other volumes covering several of the Central and East European countries also tend to approach them on an individual

6 *Theoretical and methodological considerations*

rather than on a collective level. Notably, some of these volumes hence contrast between the old and the new allies. The works of Tom Lansford and Blagovest Tashev (2005) and Robert Czulda and Marek Madej (2015) are examples.

Defence reform and military transformation covering some of the Central and East European countries are the most frequent approach. Anthony Forster and his colleagues (2002) focused, for example, specifically on post-communist Europe. Other works have included Central and East European countries alongside states from other regions when focusing on the phenomenon of military transformation rather than on strategy. Timothy Edmunds and Marjan Malesic (2005) and Thomas Bruneau and Harold Trinkunas (2008) are both examples of this. Finally, Nora Vanaga and Toms Rostoks (2019) focused on one of the aspects addressed in this book, deterring neighbouring Russia. However, in their edited volume they include not only Central and East European countries but other states as well. Moreover, they address their cases individually and, in addition, not as part of studying the military strategies of former Eastern states becoming the new allies of both NATO and the EU. Consequently, we argue that our project is unique, being a monograph focusing on the defence strategy of the 11 new allies as a group in order to explain their strategic preferences.

The aim of our project is to contribute to previous research empirically, theoretically and methodologically. Empirically, we strive to present and analyse the defence strategies of each of the 11 new allies by conducting a structured focused comparison of official strategic documents during the two first decades of the twenty-first century from each of the 11 cases. By doing this we draw attention to two categories of states, middle powers and small states, overlooked in previous research focusing on the military strategies of the great powers. Theoretically and methodologically, we aim to contribute to previous research on strategy and defence transformation processes by introducing an analytical framework to compare and explain similarities and differences in states' strategic priorities. This framework is a contribution to previous research in three main ways. First, it offers an alternative to approaches based on the assumption of states being like units, an assumption common to both structural realism and strategic research on game theory that effectively prevents research on the diversity of strategies actually pursued by states. Second, it contributes to previous research within Strategic Studies by presenting a menu of alternative alignment strategies and an operationalisation of each of the three key elements of military strategy – ends, means and ways – hereby introducing an analytical framework for systematic comparisons of defence strategies. Third, the three potential intervening variables offers a new comprehensive tool for analysing how differences related to national characteristics affect states' strategic priorities.

Our choice of intervening variables reflects an eclectic approach that combines several different theoretical traditions. The first potential intervening variable, *relative power*, relates to research on small states and middle powers. A basic premise in this field of research is that states' strategic choices are affected by power asymmetries between more or less resourceful states. Our analysis of the influence of differences related to relative power departs from a distinction between minor middle powers and small states. For reasons elaborated on in the next chapter, the Czech Republic (Czechia), Hungary, Poland and Romania are categorised as being minor middle powers, and the other seven states are categorised as small states. Researchers on the alignment and military strategies of small states have argued that these states are more sensitive to changes in the balance of power and more dependent on support from other states or institutions compared to states that are more resourceful. The political leadership of small states are also assumed to be more concerned with the survival of their own state and to concentrate their resources on short-term and local matters. Their dependency on other actors or institutions forces small states to adjust quickly to changes in their external environment, such as the breakdown of systems for collective security, increased tension between great powers and unfavourable changes in the distribution of power between the main competing regional or global great powers (Rothstein 1968; Elgström 2000; Edström et al. 2019). Research on both small states and middle powers emphasises similar preferences for institutionalised multilateral cooperation to compensate for the lack of nationally controlled resources (Edström et al. 2019; Cooper 1997). However, greater access to latent economic and military power resources and greater international reputation may provide middle powers with strategic opportunities that are not open to small states. Consequently, we expect that middle powers perceive themselves to be more dependent on external efforts than great powers are but less dependent than small powers. We would also expect that middle powers have greater ambitions regarding their internal efforts than small states but lower ambitions in comparison to great powers.

The second intervening variable, *geographical characteristics*, is related to classical realism, Strategic Studies and research on strategic culture. This variable analyses the impact of national geographical characteristics on strategic priorities. According to Colin Gray, geographical factors permeate military strategic thinking and contribute to a nation's strategic culture (Gray 2006:137–146). For reasons discussed further in the next chapter, we will concentrate our efforts to one factor related to geography: a shared land border with Russia. The change in the regional security environment that occurred after Russia's armed attack against Georgia in 2008 and Ukraine in 2014 can be assumed to be most alarming to states with land borders to Russia. If this variable proves to be important, we expect that the four states that share a land border with Russia (Estonia, Latvia, Lithuania and Poland) will be more reluctant to dismantle military resources related to national defence, especially land forces. We also expect that these four states will give greater priority to national security and respond more firmly to deteriorating regional security.

Our third intervening variable, *historical experiences*, establishes ties to both classical realism and research on strategic culture.[6] This variable rests on the assumption that strategic choices are partly based on collective interpretations of previous experiences of armed conflicts or other formative historical experiences of a particular state. In his seminal work, *Perception and Misperception in International Politics,* Robert Jervis observed that war experiences are not only important formative periods but also often such powerful experiences that lessons can be passed down to those who did not experience them directly themselves (Jervis 1976). Regarding historical lessons, we should, according to Dan Reiter, expect that 'continuity of policy follows success, while innovation follows failure'. Moreover, 'systemic wars' among great powers were assumed to be 'the most likely of wars to be formative because they are generally the most earth shaking events in world politics'. WWI and WWII were, Reiter argued, the most important systematic wars in the modern era in this regard (Reiter 1994:490 and 497).

Regarding experiences during great power wars, we will focus our attention on experiences during WWII related to three different dimensions. First, experiences of national defence against aggression; second, experience of offensive warfare in alliance with Nazi-Germany and/or the USSR; third, receiving support from one or more Western great powers. States having positive experiences in defending themselves against armed attack from another state are expected to be more reluctant to dismantle their capacities for national defence and transform their armed forces towards expeditionary warfare. States with bad experiences of national defence after suffering an armed aggression are expected to give greater priority to efforts related to collective self-defence, especially if they consider themselves threatened by a more resourceful state. The second and third dimensions concern relations to specific great powers, regarding both armed aggression and support. Some of our 11 cases were invaded and/or occupied by Nazi-Germany, others by the USSR, and some had experiences of offensive warfare together with either or both of them. Are states that were invaded and/or occupied by the USSR responding more harshly to a renewed potential Russian military threat compared to states that were not fighting the USSR? Do states that suffered attacks from Nazi-Germany find it more difficult to cooperate with the EU and/or NATO in matters related to collective defence, and are states with positive or negative experiences from joining forces with great powers more or less inclined to commit themselves to allied war efforts?

Historical experiences relating to wars and other formative experiences during state-building processes have been identified as important sources of national strategic cultures (Lantis and Howlett 2016). Obviously, the 11 new allies have different experiences from potential formative periods, such as independence or not during the Middle Ages and civil war or not when gaining independence after the end of the Cold War. Regarding the formative experiences of state-building, we use a main dividing line related to the time of *lasting* independence. The category of relatively young states, that is the states that gained their current statehood only after the end of the Cold War, are assumed to give priority to ends related to protecting their independence. The group of more mature states, that is Bulgaria, Hungary, Poland and Romania as well as Czechia as the heir of Czechoslovakia, are expected to prioritise influencing and improving their position in the international system. This may also affect their priorities regarding means, with the first category focusing on national defence and the latter developing means to be able to participate in multilateral international operations order to gain status and/or influence.

As previously mentioned, the 11 new allies share a number of relevant unit-level characteristics that may produce common challenges and generate similarities in their responses to systematic pressures. For instance, their common experiences of communist totalitarian rule, the dependency of Cold War Soviet military equipment and (for most states) common war planning and war preparations within the context of the WP creates common challenges when the new allies face pressures to adjust to NATO standards and demands for new military capabilities. Moreover, they all share the experience of being part of a multinational federation into which their national defence strategies were subordinated. Additionally, as postcommunist states, they share common challenges in adjusting to Western liberal values fundamental to both the EU and NATO. However, they also differ in some potentially relevant national characteristics. To handle this mix of similarities and differences analytically, this study employs a modified most-similar system design.[7] Our use of intervening variables deviates from similar traditional case design because the relevant variation in the explanatory variables are not found in the independent variables but in the three *intervening* variables of relative power, geography and historical experiences. Therefore, the impact of our independent variables, membership in the EU and NATO and changes in the common security environment, is assumed to be filtered by the intervening variables.[8]

The link between the intervening variables and the dependent variable of defence strategies, is investigated primarily through a cross-case analysis focusing on covariation across cases concerning the presence or absence of the three intervening variables and the outcome on the dependent variable. The purpose of this is to determine whether variation in the intervening variables coincides with variation in the dependent variable – the defence strategies – or not. In the presentation of our empirical findings, we use the method of structured focused comparison (SFC).[9] Each case study will be divided into six subsections: (1) historical background, (2) assessments of the strategic environment and the strategic responses of each country regarding (3) ends, (4) means and (5) ways. Finally, each case study contains (6) a summary in which our findings on each individual case is related to previous research on the specific state.

Each country-specific section in a chapter also presents a detailed presentation of the sources used in our analysis. The empirical material for the analysis has been retrieved solely from primary official sources. The bulk of the primary sources used in this study consists of defence white papers, national strategies and strategic defence reviews. Additionally, we used documents such as programs for the modernisation of the armed forces, military strategies and doctrines as well as defence action plans. Our choice to rely primarily on official documents is motivated by the ambition to obtain comparable systematic data possible to compare across cases. Gary King and his colleagues have in a similar manner argued that SFC is mainly characterised by its highly systematic data collection, and that the same information should be collected about the same variables in all cases with the guidance of relevant theory (King *et al.* 1994:45–46).

Obviously, certain confidential aspects of defence planning and military strategy cannot be found in public documents or be included in this study. Arguably,

10 Theoretical and methodological considerations

there is an additional risk that officially published strategic documents are declaratory rather than operational in their nature (see, for example, Edström 2003). However, in democracies, the kind of official documents used in this study serve as political guidance and direction for the development of the armed forces. Consequently, their internal role in expressing and determining strategic priorities would be lost if they were mainly declaratory. On the other hand, official documents also have the external role of communicating ambitions and intentions to other states. A part of this signalling concerns the willingness of the new European allies to live up to expectations and promises related to their membership in the EU and NATO. In this context, scepticism about whether or not a specific state has put deeds behind its words is necessary. To handle this latter problem, we have continuously challenged claims regarding ambitions and military capacities presented in the national official documents and complemented these documents with assessments on each state's military capacities published by the International Institute for Strategic Studies (IISS). Notably, the quantitative data on developments regarding military equipment collected from the IISS is not used to measure the individual states' military strength *per se*. Instead, the IISS assessments are used to analyse the extent to which the new allies *de facto* have modernised and transformed their armed forces along the lines expected by the EU and NATO and evaluate the extent to which the official ambitions of 11 new allies is matched by the overall development of their armed forces. Moreover, in the concluding part of each country's specific chapter, our own findings are related to views presented in previous research. To further improve the validity of our interpretations, we have also presented a draft version of the empirical chapters to each of the respective countries' defence attachés accredited to Sweden, along with a copy to the Swedish ditto to each of the 11 countries.

In the next chapter in this initial Part I, Chapter 2, the analytical framework of this study is further discussed and related to previous research on both our independent variable (systematic pressures related to changes in the global and regional security environments and to the memberships in the EU and NATO) and our three intervening variables (relative power, geography and historical experiences, respectively). In Chapter 3, the constitutive elements of our dependent variable, *Defence strategy*, is operationalised. Defence strategy is defined as interconnected ideas on how politically defined strategic ends should be achieved through a combination of (1) interacting, on a political level, with other states and (2) suitable strategic ways of developing and employing military means. While the former part is defined as *Alignment strategy*, the latter is defined as *Military strategy*.

The empirical analysis is presented in Part II, that is Chapters 4–14. In order to avoid unintentional clustering, the 11 new allies are addressed in alphabetical order. Moreover, the cases are presented in a similar way. Initially, the primary sources used for the analysis are introduced. Then the historical background is illuminated. This is followed by four sections, analysing each of the elements of the dependent variable. In the last section, we present our conclusions and relate our results to the findings of previous research.

In Part III, the conclusions of the exploration are summarised. In Chapter 15, the results from the empirical exploration are aggregated. Based on the findings, the explanatory power of each of the intervening variables is tested in Chapter 16. Finally, our overarching conclusions as well as the answers to the overarching question of this study are presented in Chapter 17.

Notes

1 In 1992, the Maastricht Treaty transformed the EC into the EU.
2 In 1995, three additional states (Austria, Finland and Sweden) also gained membership.
3 Other important events leading to the end of the Cold War include (1) the German unification in October 1990, when former communist German Democratic Republic (GDR or East Germany), became a part of the Federal Republic of Germany (FRG or West Germany), and (2) the dissolution of the WP in February 1991.
4 Albania, Montenegro and North Macedonia are excluded from the analysis since these three countries, as of August 2021, were members of NATO.
5 The concept of strategic culture was first articulated during the Cold War when an American political scientist, Jack Snyder, observed that the USSR approached strategy in a rather different manner compared to the US and other Western states (Snyder 1977). Snyder soon gained support from other scholars, such as Colin Gray, who claimed the existence of a unique American strategic culture (Gray 1981). In the following decades, the concept has been used in many different ways, and it remains subject to scholarly debate. See, for instance, Johnston (1995a, 1995b), Kier (1997), Gray (1999), Neumann and Heikka (2005).
6 For a discussion on the possibility of combining assumptions associated with realism and research on strategic culture, see Rynning (2011).
7 Comparative methods derive their basic logic from John Stuart Mill's method of agreement and method of difference. Depending on the author, different names are used to refer to these logics. For example, the method of agreement is sometimes referred to as the most-different systems design or the positive comparative method. The method of difference is also called the most-similar systems design or the negative comparative method (George and Bennet 2005).
8 According to Alexander George and Andrew Bennet (2005), intervening variables are used to explain cases with multiple interaction effects. The intervening variables are found between the cause and effect variables and constitute parts of a causal mechanism. For other examples of research on strategy using unit-level characteristics as intervening variables, see Rose (1998), Schweller (2004) and Rathbun (2008).
9 The research method SFC relies on two major components. First, its 'structure' is borrowed from statistical or survey methods and implies a reliance on 'asking a set of standardized, general questions to each case'. These questions should reflect the research objective and theoretical focus and is intended to produce comparable data that allows 'cumulative development of knowledge' about a given phenomenon. (George and Bennett 2005:69–70).

Bibliography

Booth, Ken (1990). 'The Concept of Strategic Culture Confirmed' in Carl Jacobsen (ed). *Strategic Power: USA/USSR*. Basingstoke: Macmillan.
Bruneau, Thomas and Harold Trinkunas (eds) (2008). *Global Politics of Defense Reform*. Basingstoke: Palgrave Macmillan.

12 Theoretical and methodological considerations

Churchill, Winston (2003). *Never Give in! The Best of Winston Churchill's Speeches* (Selected by His Grandson, Winston S. Churchill). New York: Hyperion.

Cooper, Andrew (1997). 'Niche Diplomacy: A Conceptual Overview' in Andrew Cooper (ed). *Niche Diplomacy: Middle Powers after the Cold War*. London: Macmillan Press.

Czulda, Robert and Marek Madej (eds) (2015). *Newcomers No More? Contemporary NATO and the Future of the Enlargement from the Perspective of "Post-Cold War" Members*. Warsaw: International Relations Research Institute.

Edmunds, Timothy and Marjan Malesic (eds) (2005). *Defence Transformation in Europe: Evolving Military Roles*. Amsterdam: IOS Press.

Edström, Håkan (2003). *Hur styrs Försvarsmakten? – Politisk och militär syn på försvarsdoktrin under 1990-talet*. Umeå: Umeå University.

Edström, Håkan, Dennis Gyllensporre and Jacob Westberg (2019). *Military Strategy of Small States: Responding to the External Shocks of the 21st Century*. Abingdon: Routledge.

Elgström, Ole. (2000). *Images and Strategies for Autonomy: Explaining Swedish Security Policy Strategies in the 19th Century*. London: Kluwer Academic Publishers.

Forster, Anthony, Timothy Edmunds and Andrew Cottey (eds) (2002). *The Challenge of Military Reform in Post-Communist Europe*. Basingstoke: Palgrave Macmillan.

George, Alexander and Andrew Bennet (2005). *Case Studies and Theory Development in the Social Sciences*. Cambridge, MA: MIT Press.

Gray, Colin (1981). 'National Style in Strategy: The American Example' *International Security* Volume 6, Issue 2.

——— (1999). *Modern Strategy*. Oxford: Oxford University Press.

——— (2006). *Strategy and History: Essays on Theory and Practice*. Abingdon: Routledge.

Hadžić, Miroslav, Milorad Timotić and Predrag Petrović (eds) (2010). *Security Policies in the Western Balkans*. Belgrade: Belgrade Centre for Security Policy.

Jervis, Robert (1976). *Perception and Misperception in International Politics*. Princeton, NJ: Princeton University Press.

Johnston, Alastair Iain (1995a). *Cultural Realism: Strategic Culture and Grand Strategy in Ming China*. Princeton, NJ: Princeton University Press.

——— (1995b). 'Thinking about Strategic Culture' *International Security* Volume 19, Issue 4.

Kier, Elizabeth (1997). *Imagining War: French and British Military Doctrine between the Wars*. Princeton, NJ: Princeton University Press.

King, Gary, Robert Keohane and Sidney Verba (1994). *Designing Social Inquiry: Scientific Inference in Qualitative Research*. Princeton, NJ: Princeton University Press.

Lansford, Tom and Blagovest Tashev (eds) (2005). *Old Europe, New Europe and the US: Renegotiating Transatlantic Security in the Post 9/11 Era*. Abingdon: Routledge.

Lantis, John and Darryl Howlett (2016). 'Strategic Culture' in John Baylis, James Wirtz and Colin Gray (eds). *Strategy in the Contemporary World*. Oxford: Oxford University Press.

Matlary, Janne Haaland (2009). *European Union Security Dynamics: In the New National Interest*. Basingstoke: Palgrave Macmillan.

Neumann, Iver and Henrikke Heikka (2005). 'Grand Strategy, Strategic Culture, Practice: The Social Roots of Nordic Defence' *Cooperation and Security* Volume 40, Issue 1.

Péczeli, Anna (ed) (2019). *The Relations of Central European Countries with the United States*. Budapest: Dialóg Campus.

Rathbun, Brian (2008). 'A Rose by Any Other Name: Neoclassical Realism as the Logical and Necessary Extension of Structural Realism' *Security Studies* Volume 17, Issue 2.

Reiter, Dan (1994). 'Learning, Realism, and Alliances: The Weight of the Shadow of the Past' *World Politics* Volume 46, Issue 4.

Rose, Gideon (1998). 'Neoclassical Realism and Theories of Foreign Policy' *World Politics* Volume 51, Issue 1.

Rothstein, Robert (1968). *Alliances and Small Powers*. New York: Columbia University Press.

Russian Ministry of Foreign Affairs (MoFA) (2021a). *Agreement on Measures to Ensure the Security of the Russian Federation and Member States of the North Atlantic Treaty Organization*.

——— (2021b). *Treaty between the United States of America and the Russian Federation on Security Guarantees*.

Rynning, Sten (2011). 'Strategic Culture and the Common Security and Defence Policy: A Classic Realist Assessment and Critique' *Contemporary Security Policy* Volume 32, Issue 3.

Schweller, Randall (2004). 'Unanswered Threats: A Neoclassical Realist Theory of Underbalancing' *International Security* Volume 29, Issue 2.

Snyder, Jack (1977). *The Soviet Strategic Culture: Implications for Limited Nuclear Operations*. Santa Monica, CA: RAND.

Vanaga, Nora and Toms Rostoks (eds) (2019). *Deterring Russia in Europe: Defence Strategies for Neighbouring States*. Abingdon: Routledge.

Zajac, Justyna (2016). *Poland's Security Policy: The West, Russia, and the Changing International Order*. Basingstoke: Palgrave Macmillan.

2 Analysing and explaining strategic adjustment and diversity

> Yalta did not ratify a natural divide, it divided a living civilization. The partition of Europe was not a fact of geography, it was an act of violence. And wise leaders for decades have found the hope of European peace in the hope of greater unity. [...] Consider how far we have come since that speech. Through trenches and shell-fire, through death camps and bombed-out cities, through gulags and food lines men and women have dreamed of what my father called a Europe "whole and free." This free Europe is no longer a dream. It is the Europe that is rising around us. [...] All of Europe's new democracies, from the Baltic to the Black Sea and all that lie between, should have the same chance for security and freedom – and the same chance to join the institutions of Europe – as Europe's old democracies have.
>
> George W. Bush (2001)

This chapter introduces both the independent and the intervening variables we use in the analysis of the defence strategies of the new European allies, the 11 former communist states that have become members of both the European Union (EU) and the North Atlantic Treaty Organization (NATO). The chapter has two main sections. The first section focuses on the independent variable, the expected consequences of membership and common challenges related to changes in the external environment during the two first decades of the twenty-first century. The second section introduces the three intervening variables that are used to explain differences in strategic choices among our cases. The three variables are (1) differences in relative power and position in the international system, (2) national geographical characteristics and (3) historical experiences related to formative periods of state-building processes as well as of wars and armed conflicts.

2.1 The new and old European security dynamics

During the Cold War era, the power competition between the United States (US) and the Soviet Union (USSR) affected and overlaid the internal security dynamics in almost all regions (Buzan and Wæver 2003). The divided Europe was at the centre of this competition. With the creation of NATO and the Warsaw Pact (WP) in 1949 and 1955, respectively, a Western and Eastern military bloc were formalised, leaving only a small group of military non-aligned or neutral states on the sidelines. Between 1950 and 1952, six continental Western European states –France, the Federal Republic of Germany (FRG or West Germany), Italy and the three BENELUX countries, that is Belgium, the Netherlands and Luxembourg – negotiated and established a complementary security cooperation, the European Coal and Steel Community (ECSC). ECSC was presented as a first step towards an even closer union that ultimately aimed at preventing new wars between European great powers by establishing common supranational institutions and creating closer ties among its citizens and member states. This development continued with the establishment of European Economic Community (EEC), which later on developed into the European Communities (EC). During the 1970s and 1980s the EC attracted and allowed six additional European states to become members. With the exception of Ireland, these states were also members of NATO.

The implosion of the USSR and the dissolution of the WP put an end to the bipolar East–West division of Europe and left the US as the sole superpower in a new global and US-led unipolar order. In Europe, old ideas and promises of a peaceful, prosperous and united Europe were perceived as both an opportunity and an obligation that Western Europe had to the formerly suppressed Central and East European states. The EU and NATO both had central roles in offering institutional platforms for overcoming the previous bipolar division of Europe. In 1995 three formerly non-aligned states (Austria, Finland and Sweden) joined the EU. The 11 new allies considered membership in these two organisations as essential to their future prosperity and security. However, membership came with a price. To gain membership, new members have to fulfil a number of explicit criteria. In addition, membership creates pressures to adjust to common norms and contribute to common efforts. In this chapter, we will briefly present the different membership criteria of the EU and NATO. However, our focus will be on general systematic pressures related to changes in the overall security dynamics in Europe as well as to defence transformation processes within the EU and NATO.

2.1.1 The new post-national security dynamics

Reflecting on the new security dynamics in Europe in the post–Cold War era, Janne Haaland Matlary, identified a dramatic change regarding prevailing ideas on the character of war and the use of force. In a book published in 2009, Matlary argued that the 'nation-state model of defence' – the view of wars as existential struggles between competing nation-states, capable of mobilising the total material and human resources of a society and threatening to extinguish each other – has been replaced with new 'post-national' ideas regarding of the use of armed force. According to this view, states engage in 'optional wars' using their military capacities far away from their own borders creating security for 'strangers' together with other states rather than defending their own territory and citizens. Consequently, the notion of national territorial defence is becoming 'less and less relevant for the organisation, definition and actual use of military force' and the security policies of European countries are both 'de-territorialised' and 'de-nationalised' (Matlary 2009:3 and 24–25).

Then Norwegian Chief of Defence Sverre Diesen argued in a similar manner. He claimed that the fall of the 'totalitarian ideologies' and the political and economic integration in 'our part of the world', meant that we no longer had to calculate with the risk of new existential wars between states. Moreover, since this situation was the result of 'irreversible' processes, the whole idea of existential wars could safely be put on the 'scrap-heap of history'. According to Diesen, this development also meant the end of the general view of defence and warfare associated with the 'Napoleonic paradigm' and the birth of a new defence paradigm. The armed forces of the Napoleonic paradigm had their prime use in fighting 'total wars', and their strength was based on successful nation-building processes, social integration and the ability to use conscription to mobilise a large part of their citizens in the nation's war efforts. The armed forces of the new paradigm, on the other hand, are used in limited conflicts to achieve limited political purposes and are manned by standing professional soldiers motivated by professional values and cultures (Diesen 2005:167–170).

Diesen was not the only top-ranking officer who reflected on the changing character of war during the first decade of the twenty-first century. In a celebrated study from 2007, British General Rubert Smith argued that the 'old paradigm of interstate industrial wars' had ended and been replaced with a new paradigm: 'war amongst the people'. According to Smith, the last battle in which tanks in formation was the deciding force, took place in the 1973 Arab–Israeli war, and war conceived as 'battle in a field between man and machinery' and 'a massive deciding event in a dispute in international affairs [. . .] no longer exists' (Smith 2007:3–5). The new paradigm is, Smith argued, characterised by more 'malleable objectives' related to individuals and societies that are not states. Due to intense media coverage, the fighting occurs in every living room as well as on the streets and fields of a conflict zone. Moreover, war efforts are guided by the ambition of avoiding casualties rather than using armed force to achieve prioritised national war aims at any cost. Additionally, the confrontations of the twenty-first century are in most cases conducted in some form of multilateral grouping rather than of national armies fighting one another (Smith 2007:19).

In Europe and other parts of the Western world, security considerations related to the bipolar tension and competition between East and West were superseded by considerations related to more diffuse risk and challenges entailing a variety of possible uses of military force (Rasmussen 2006). These new threat perceptions often concerned non-state actors such as organised international crime and international terrorism, the spread of weapons of mass destruction and civil wars fuelled by domestic ethnic or religious conflicts and scarce resources between sub-state actors. The wars in the Balkans and other conflicts in Africa and Asia created new demands on the armed forces of states participating in various operations related to crisis management, peacekeeping and peace enforcement. According to Adrian Hyde-Price, all European countries struggled throughout the 1990s to understand and come to terms with the implications of these changes in the nature of international conflicts and the potential use of armed force (Hyde-Price 2004:331).

States whose political leadership shared this vision of a new era in European or even global history initiated defence transformation processes with severe consequences for their armed forces' capability to conduct independent defence of their own territory from an attack from other states. In such states, national defence as a policy area experienced a 'de-securitization' and normalisation, meaning that issues related to national defence lost its special standing and was treated as any other policy area (Buzan and Wæver 2003). One effect of this was shrinking defence budgets. At the same time, participation in international military operations demanded that the armed forces of these states developed new capabilities for multilateral expeditionary warfare involving both new professional skills, new tactics and new or modified military equipment and weapon systems. To be able to develop these new capabilities with no extra funding, most states had to make drastic cuts in the size of their armed forces. Deepened multilateral cooperation related to both the use of force and force generation was an additional key instrument in these defence transformation processes.[1]

In an analysis of the new dynamics multilateral defence cooperation, Matlary identified a number of different motives for deepened cooperation supposedly common to all Western states. Since the new wars are 'optional' and do not directly concern the defence of the contributing states' own territory, governments must place premium value on force protection. This in turn induces government to seek 'multilateral risk-sharing' by acting in operation areas together with armed forces from many different states. Additionally, being a part of a coalition with a mandate from an international organisation creates opportunities to transfer blame and responsibility to the organisation in question, and operations pursued in the name of international organisations such as the EU, NATO and the United Nations (UN) are easier to legitimate. Moreover, shrinking defence budgets and explosive cost increases in various advanced weapon systems prevents an increasing number of states from keeping balanced defence structures with a complete catalogue of military capabilities. Even governments representing relatively resourceful states depend increasingly on multilateralism in procurement, maintenance and deployment. Furthermore, European small states that would otherwise have very limited capacities for unilateral action on the world stage may use the EU as a 'force multiplier' and use participation in peace support operations (PSOs) as a way of gaining status and political influence (Matlary 2009:7–8).

According to Timothy Edmunds, political and military integration across Europe in 2005 had already made it increasingly problematic to divide the challenges faced by the armed forces in Europe along the old East–West lines. Even so, the former communist states still share some common 'themes and issues' such as 'particular organisational legacies of communist-era mass armies'. However, the defence transformational challenges that they are facing is converging with the challenges the EU and NATO's older member states are facing. A first such common challenge concerned the previously discussed changing nature of 'the defence of national territory imperative'. During the Cold War, this imperative provided a rationale for the armed forces force structure, budget and legitimacy. The lack of existential threats and an increased emphasis on threats from non-state actors have made this imperative less central to 'most European states'. However, Edmunds argued, the role of the military as the 'ultimate guarantor of states' security has not disappeared and is not likely to do so any time soon. Defence reviews across Europe retain interpretations of national defence as the 'foundational justification' for their armed forces, and the national territorial imperative still has strong support both within the armed forces themselves and in European societies. Moreover, the defence procurement strategies of many European states have continued to focus on high-value equipment such as fighter aircraft, predicated specifically on territorial defence despite their 'questionable utility in the new European security environment' and the difficulty states have in deploying and using them effectively (Edmunds 2005:9–10).

Nevertheless, Edmunds argued, the post–Cold War era is characterised by a 'significant shift' in the kind of missions that the European armed forces are expected to carry out. These new missions concerned in particular contributions to multinational international operations to counter threats related to the supply of strategic resources, the proliferation of weapons of mass destruction (WMD) to hostile regimes, regional instability caused by intra-state conflicts and international terrorism. To address these challenges, states must be able to deploy military power swiftly and effectively wherever needed. Changes in military technology enable states to fulfil power projection missions more effectively but also raise the cost of military transformations processes. Additionally, ambitions to develop capacities for power projection have implications on the needed competences and recruitment processes of armed forces. These missions necessitate flexible and technologically advanced force structures and high skill levels from soldier of all ranks, which predicates a shift towards 'all volunteer professional forces – very different from the conscript-based mass armies of the Cold War period'. The new tasks for European armed forces presented themselves in the context of evolving alliance demands and commitments which, according to Edmunds, exercised important external influence on military reform programs. This was particularly the case for states involved in the NATO accession process, 'which has created unprecedented pressures' for the armed forces to develop the capabilities needed to participate in multinational military operations. Institutionally, such pressures were channelized through formalised programs as NATO's Partnership for Peace (PfP) and Membership Action Plan (MAP). Within the EU, the 1999 Helsinki Head Line Goals exercised a similar influence. Participation in PSOs provided, according to Edmunds, contributing armed forces with a new source of legitimacy for continued defence expenditures. Even if force commitments from small states were modest in size, they could still be presented as evidence of commitments to the wider goals and activities of the alliance (Edmunds 2005:10–11 and 14).

The formal criteria for eligibility to NATO membership in the post–Cold War era were initially outlined in the 1995 NATO report *Study on NATO Enlargement*. The report stressed that the goal of the enlargement was to 'render obsolete the idea of "dividing lines" in Europe'. According to this document, the eligibility criteria included a functioning democratic system and market economy, democratic civil–military relations, treatment of ethnic minorities in accordance with guidelines of the Organization for Security and Co-operation in Europe (OSCE), the resolution of all disputes with neighbours and a general commitment to the peaceful settlement of disputes. Moreover, candidate countries should have the ability and willingness to give military contributions to the alliance and achieve interoperability with allied armed forces. PfP exercises and the Planning and Review Process (PARP) would introduce partners to collective defence planning and pave the way for more detailed operational planning (Drent *et al.* 2001:5; Simon 2001). In 1999, NATO's Washington summit introduced the MAP as a 'road map' for applicant states to develop military capabilities compatible with NATO's operational capabilities concept. The MAP requires the submission of tailored Annual National Plans covering various political, economic, legal and defence-related aspects of membership. Previous participation in WP defence planning had, according to Jeffrey Simon, left many of the new allies with decaying Soviet technology and an oversized force structure, too big and too heavy for NATO's present needs (Simon 2001:29–31).

In the conclusions from the meeting of the European Council in Copenhagen in June 1993, the council announced that the associated countries in 'Central and Eastern Europe that so desire shall become members of the European Union'. The conclusion also specified a number of economic and political conditions for membership, the so-called Copenhagen criteria. These criteria partly overlap with some of the criteria in the NATO study from 1995. However, the EU criteria do not concern military capacities or civil–military relations specifically. According to the Copenhagen criteria, membership requires that the 'candidate country has achieved stability of institutions guaranteeing democracy, the rule of law, human rights and respect for and protection of minorities'. Moreover, membership required a 'functioning market economy able to cope with competitive pressure and market forces within the Union' and 'ability to take on the obligations of membership including adherence to the aims of political, economic and monetary union' (European Council 1993). The criteria presented by the EU made accession conditional on the candidate countries' ability to fulfil the membership criteria, and the EU established a screening process led by the Commission of the EU.

Research on Europeanisation processes have identified a number of different mechanisms that can explain how common institutions affect the policies of member states. According to Kyriakos Moumoutzis and Sotirios Zartaloudis, 'policy learning' has been the most common explanation of policy change as the outcome of Europeanization. Policy learning is defined as 'relatively enduring changes in thought or behavioural intention that result from experience and/or new information concerned with the attainment or revision of policy objectives'. Moumoutzis and Zartaloudis distinguished between two distinct types of learning, 'instrumental learning' that concerns new and more cost-effective use of means and 'social learning' that concerned a change in policy ends and objectives. Instrumental learning is often used in rationalistic explanations assuming that actors choose the policy alternative that produces a preferred outcome at the lowest possible cost.

Social learning, on the other hand, is based on the assumption that actors adjust to common rules and norms in the ambition of living up to common expectations. This second type of learning argues that actors may change their identities, basic preferences and ends to fit into a new European context (Moumoutzis and Zartaloudis 2016:340–343).

'Socialization' has been presented as an alternative concept to social learning. According to Nicole Alecu de Flers and Patrick Müller, socialization can be understood as a process whereby 'actors of a given community are inducted into the community's rules, norms and policy paradigms' (de Flers and Müller 2012:24). However, the extent to which individual states adjust their policies or preferences is, according to de Flers and Müller, likely to differ among individual member states due to domestic factors such as the size of the member state and its foreign relation network. Larger member states are more frequently portrayed as 'shapers' rather than as 'takers' of common European policies, and the 'EU impact' on smaller member states is 'considered usually to be more profound'. Other domestic variables that may produce a diversity in strategic responses are 'historically conditioned variables', such as national identity, for example an 'Atlanticist' versus an 'Europeanist' orientation, and differences related to strategic culture, for example views concerning the use of force. An additional unit-level characteristic mentioned by de Flers and Müller is the duration of EU membership. In contrast to old member states, new member states have not had the opportunity to influence the EU foreign policy *acquis* from the outset of the cooperation, and they therefore had to adjust their national policies to pre-established European positions. Nevertheless, de Flers and Müller also cautioned against overstating the EU's impact on national policies. Europeanization processes may be reversible and challenged by processes of 'de-Europeanization' or 'renationalization' as member states 'fall back on their own resources and individual strategies during political crises or changes in government' (de Flers and Müller 2012:23).

The new security dynamics discussed so far have potentially far-reaching effects on individual states' military strategies. Regarding *ends*, the new security dynamics suggest a move away from prioritising goals related to national security or survival and toward prioritising goals related to influence and status. Regarding *means*, the new dynamics suggests that even less resourceful states will increase their efforts to develop capacities for expeditionary warfare, making them cable of contributing to PSOs or coalitions of the willing led by more resourceful states. Regarding *ways*, the new dynamics suggest a decreased emphasis in unilateral methods and an increased emphasis on multilateral methods (for a more detailed discussion on these three elements of military strategy, see Chapter 3). However, as discussed in Section 2.2, national characteristics related to relative power, geography and historical experiences may make states more or less sensitive to these common systematic pressures and processes of learning and socialization. Governments may disagree with the assumptions of a new stable and peaceful European security order and continue to perceive that their state is exposed to threats from other states. Such states are less likely to follow the path of the new security dynamics. Other states may find that their own lack of resources do not allow them to develop military capacities related to expeditionary warfare.

2.1.2 Return of the old European security dynamics

The reluctance of some states to adjust to the new security dynamics may also be explained by changes in the regional pattern of interactions and war expectations from 2008 onwards. In a previous study on Denmark's, Finland's, Norway's and Sweden's responses to Russia's military aggression against Georgia in 2008 and Ukraine in 2014, we noticed that these states adjusted their strategies to a perceived deteriorating regional security environment and changing priorities within the EU and NATO. Although their responses differed, there was an overall increased emphasis on the ends and means related to national and collective defence, and the possibility of new interstate wars involving member states of the EU and NATO was back as a main security concern (Edström et al. 2019; Edström and Westberg 2020b).

In the terminology of Regional Security Complex Theory (RSCT), the pattern of 'amity and enmity' in Europe was changing towards increased war expectations (Buzan and Wæver 2003). Obviously, these expectations did not concern an increased risk of wars among the members of the EU security community. Instead, Russia's demonstrated ability and will to use military force for political purposes, its changed patterns of military exercises, violations of other states' rights to territorial integrity and open challenge of fundamental principles of the European post–Cold War security order indicate that the long era of interstate warfare in Europe has not come to an end. NATO responded to this challenge by activating its defence planning, increasing the number and scope of military exercises in exposed regions and enhancing efforts for a forward presence in Central and Northern Europe by circulating military units from less exposed member states to states with land borders to Russia. However, parallel to these efforts the US has voiced demands for an increased burden sharing among the US and the European NATO member states and partners. The EU, on the other hand, has so far not been able to move forward with its ambitions regarding collective defence and collective security mentioned in Article 42 of the Treaty of Lisbon. An additional challenge to the member states of the EU and NATO is that transnational threats related to non-state actors and failed states still has to be addressed, creating potential institutional pressures for maintaining capacities for expeditionary warfare (Edström and Westberg 2020a, 2021).

How have the 11 new allies responded to the institutional pressures related to the membership process and the changes in the European security dynamics during the twenty-first century, and are their responses similar or diverging? Their relatively recent war experiences from the Second World War (WWII) and their common experiences during the Cold War may suggest that we should expect similarities in their strategic responses and common challenges related to their previous dependence on Soviet military equipment and WP defence planning. However, their strategic responses to the changes in their external security environment during the first two decades of the twenty-first century have been far from uniform. In the next section, we discuss how three intervening variables related to differences in relative power and position in the international system, geographical location and historical experiences can be used to explain differences in the defence strategies among the 11 new allies and similarities in strategic priorities among smaller groups of countries that share one or more unit-level characteristics.

2.2 National characteristics: three intervening variables

Our choice of intervening variables reflects an eclectic approach that combines explanatory variables from several different theoretical traditions. Our first intervening variable, *relative power*, relates to research on small states and middle powers that focus on how power asymmetries between different categories of states make these states develop specific small state or middle power strategies that take differences in relative power into account (Baker Fox 1959; Vital 1967; Rothstein 1968; Holbraad 1984; Gilley and O'Neil 2014a). More specifically, we will focus on aggregated differences related to economic, military and political resources. These capabilities are measured as the size and development of each country's economy (gross domestic product [GDP] and GDP per capita), military expenditure (aggregated 10-year annual defence spending and defence expenditure as share of GDP) and diplomatic representation (number of hosted foreign embassies). In summarising these three indicators, our 11 cases are divided into two categories: minor middle powers and small states. To what extent can we see similarities in the defence strategies among states within each group when it comes to alignment and military strategies?

Our second intervening variable, *geographical characteristics*, is related to research on geopolitics and military strategy. The word 'geopolitics' was coined in 1899 by the Swedish political scientist Rudolf Kjellén to describe the geographical characteristics of a particular state, its natural endowment and resources. During the twentieth century, the concept became an integrated part of the strategies and ambitions of the leading great powers. Halford Mackinder, who began his academic career teaching geography at Oxford University, expanded the analysis of geography to include the global power competition and the interplay between transport technology and the rise of new powers (Tuathail 2006). In an article from 1904, Mackinder prophesised that an expanded railway network would allow Russia to establish itself as a dominating power in the Eurasian 'pivot area', exercising pressure on Scandinavia, Poland, Turkey, Persia, China and India (Mackinder 2006:37). In Germany, the term 'geopolitics' was taken up by a former general, Karl Haushofer, who founded the journal *Zeitschrift für Geopolitik* in 1924. The same year he published an article with the title 'Why Geopolitics?', in which he praised Mackinder's analysis and foresight and argued that Germany 'must emerge out of the narrowness of her present living space'. He also discussed his ideas with Adolf Hitler during the latter's imprisonment in 1924 (Haushofer 2006:41; Tuathail 2006:24).

Due to its association with Nazi-Germany's expansionist war agenda, the concept of geopolitics became taboo after WWII. However, the global power competition between the USSR and the US during the Cold War era had plenty of geopolitical visions closely integrated with each superpower's grand strategy. The containment policies practiced along the lines proclaimed in the Truman doctrine and the Brezhnev doctrine's call for a common responsibility of all communist parties to protect socialism both at home and in other socialist countries are examples of this (Brezhnev 2006; Truman 2006). In the post–Cold War era, Francis Fukuyama's *The End of History and the Last Man* and Samuel Huntington's *The Clash of Civilizations and the Remaking of World Order* provided two alternative conceptualisations of a new world order replacing the bipolar competition between East and West (Fukuyama 1992; Huntington 1993). George H. W. Bush's vision of 'Europe whole and free' corresponded foremost to Fukuyama's vision. Impressed with the improvement of the relationship between Russia and the Western great powers and institutions, some observers in the early years of the new millennium argued that there is a need to rethink the concept of 'the west' and that 'the north' is a more 'relevant concept in the context of the early twenty-first century' (Lieven and Trenin 2003:12–13).

During the twentieth century, the 11 new allies were mostly on the receiving end of these competing geopolitical conceptions of great power competition. However, small states and middle powers may have their own geopolitical conceptualisations, reflecting their national experiences and perceptions of great power strategies (see, for example, Drulàk 2012; Kuus 2012). The Munich Agreement in 1938, the Molotov–Ribbentrop Pact a year later and the Yalta Agreement from 1945 that acknowledged a Russian sphere of influence in Soviet Eastern and Central Europe constitute, according to Ainius Lašas, a 'black trinity' that together affected all of the 11 new allies. This black trinity also created a 'collective guilt' among Western great powers involved in the three agreements, a sense of guilt that may explain the rationale for the dual enlargements process of the EU and NATO (Lašas 2010:8–12). Russia's invasion of Ukraine on 24 February 2022 and its renewed demands for a sphere of influence encompassing former USSR republics and former members of the WP are painful reminders of the continued legacy of the black trinity.

Research on military strategy has always recognised the importance of geography in explanations of differences in defence planning between, for instance, land-based powers and sea powers, natural barriers and strategic depth as a protection against invading forces (Gray 2006:137–139; Morgenthau 2006:122–124). National geographical characteristics have also been seen as one important element in explanations of differences related to strategic culture. Due to our great number of cases, this variable will not be analysed to the fullest extent in this study. Instead, we will focus on one central geographical aspect, a shared land border with Russia. To which extent does this single geographical characteristic produce differences in the defence strategies among our cases?

The third intervening variable, *historical experiences*, is a central element in most analyses of national strategic cultures (see, for instance, Kier 1997; Gray

1999). In our analysis of differences related to historical experiences, we will focus on two key aspects of special importance for questions related to the use of force: (1) formative experiences related to state-building processes and (2) experiences during great power wars. This third variable concerns shared supposedly intersubjective understandings of the state's own place in the world and its identity and relations to other states. In clustering the 11 states on this variable, we had to use several groupings. This variable establishes ties to both classical realism and constructivist research.[2]

In the next two subsections, we elaborate on these intervening variables and relate them to previous research. In the final and third subsection, we aggregate the elaboration and introduce the phenomenon of strategic exposure.

2.2.1 Relative power and position in the international system

Research inspired by structural realism has generally focused on the alignment and military strategies of the great powers, paying little attention to how power asymmetries between states will influence their perceived interests and their choice of strategic means and ways. Moreover, both classical realists such as Hans Morgenthau and structural realists such as Waltz base their argument on the assumption that states are 'like units' with similar basic offensive or defensive interests. According to Morgenthau, 'international politics, like all politics, is a struggle for power' and interest defined as power is assumed to be 'an objective category', 'universally valid' and ultimately finding its sources in human nature (Morgenthau 2006:10 and 29). Waltz, on the other hand, argued that since there is no division of labour between the units in anarchic systems, 'the system is composed of like units' whose most fundamental interest is to protect their survival. However, he also emphasised that states are not alike in their abilities to achieve their aims (Waltz 1979:96 and 101). The assumption of states being 'like units' comes with a high price. Faithfulness to this assumption prevents scholars from analysing the diversity of real-world strategic choices made by states. Moreover, in an article from 1993, Waltz himself acknowledged that states 'will from their different historical experiences, geographical locations, and economic interest, interpret events differently and often prefer different policies' (Waltz 1993:74).

Scholars working in the neoclassical realist tradition have explained differences in strategic choices and performances by referring to differences related to the domestic systems of states.[3] Unlike some of the research within the neoclassical tradition, we will not enter the arena of domestic politics. Instead, we hold on to the classical realist core assumption of states being unitary actors, and our choice of intervening variables concerns differences related to national characteristics (Legro and Moravcsik 1999). Our first intervening variable, *relative power*, relates to research on how power asymmetries between different categories of states make states develop strategies that take these differences in relative power into account.[4] The habit of separating states into classes, such as great, middle, and small, have figured in International Relations (IR) studies for several centuries (Holbraad 1984; Wight 1986). These hierarchical conceptions of the state system often include categories such as superpowers and great powers, middle powers and small states.[5] However, there is no agreement on how these different categories shall be separated and which indicators should be used to identify a particular state category.[6] James Manicom and Jeffrey Reeves identified three different approaches to middle power research: the ideational, the behavioural, and the positional (Manicom and Reeves 2014:29). Other scholars have made similar distinctions presenting self-identification, behavioural patterns and relative power capacity as three competing ways of identifying middle powers (Cooper *et al.* 1993; Chapnick 1999; Robertson 2017). Our definition of middle powers is mainly inspired by the positional approach. The positional approach is characterised by its focus on quantifiable indicators of power asymmetries between states in terms of differences in quantifiable latent power resources such as population size, military expenditures and GDP (Carr 2014:71–72).

A problem with quantitative definitions is that the 'cut-off line', the precise chosen number for the size of the economy, population or other relevant power resources that may be used to distinguish different classes of states, is always arbitrary (Neumann and Gstöhl 2006). To avoid this problem, our indicators on power resources combine different elements of national power (economic, military and political power) and do not specify an exact number for each cut-off line. Instead, the latent power resources between and within each category are specified relative to other categories. Using a positional approach, the new allies can be categorised as being either middle powers or small states. According to James Manicom and Jeffrey Reeves (2014), states ranking roughly within the range of 10–30 on various capability indexes can be characterised as middle powers. In a previous study, we identified a subcategory of states labelled 'major middle powers'. To qualify as a major middle power, a state must meet the double criteria of being ranked among the top 20 on both GDP and accumulated military expenditure. Furthermore, to be able to act as a middle power and to develop specific middle power strategies, states must also be recognised by other states as having this particular status. Membership in the Group of Twenty (G20) was therefore used as an additional third criterion in identifying major middle powers.

None of the 11 new allies fulfils this positive 20–20–20 criterion. To be able to analyse whether differences in defence strategies covariates with differences in relative power and the states' positions in the state system, we will use a distinction between minor middle powers and small states based on criteria similar to the general criteria we used in previous studies on great powers and middle powers. According to this definition, minor middle powers are characterised by the negative criterion of not fulfilling the criterion for major middle powers. In addition, they must also fulfil the two positive criteria of (1) being among the world's top 70 largest economies in terms of GDP and GDP per capita and (2) having accumulated military expenditures over the last 10 years among the top 70 largest. Since there is no 'minor middle power club' corresponding to the great power's permanent seats in the UN Security Council (UNSC) or the G20 for major middle powers, we have instead collected data on diplomatic representation measured as the number of foreign embassies in each state. This serves as the third positive criterion: (3) to be recognised as a minor middle power, the state must host a minimum 70 of foreign embassies. Together, these three aspects establish a positive 70–70–70 criterion.

26 *Theoretical and methodological considerations*

Using data regarding the economic power resources from the International Monetary Fund (IMF), we argue that the 11 new allies can be divided into two clusters, with five states (the Czech Republic (Czechia), Hungary, Poland, Romania and Slovakia) having a clear advantage compared to the other six states (see Table 2.1).

Table 2.1 Average GDP and GDP per capita 2010–2019.

	GDP in USD billions (world rank)	GDP per capita nominal in USD
Poland	523.4 (24)	13 769
Czechia	217.1 (49)	20 590
Romania	198.1 (51)	9 985
Hungary	138.9 (59)	14 069
Slovakia	96.9 (64)	17 878
Croatia	57.2 (76)	13 590
Bulgaria	57.2 (77)	7 979
Slovenia	48.9 (84)	23 736
Lithuania	45.8 (89)	15 765
Latvia	29.6 (102)	14 821
Estonia	25.5 (103)	19 293

Source: IMF (2021).

Data regarding military power resources has been collected from the Stockholm International Peace Research Institute (SIPRI) and is summarized in Table 2.2. So far, our indicators on latent power correlates, suggesting that five states can be considered as potential minor middle powers. However, Poland has a great advantage over the other states in both economic and military terms. Arguably, this may mean that Poland should be classified as a 'middle power' rather than as a 'minor middle power'. However, since Poland is the only state among the new allies aspiring for an upgrading, we decided to initially include Poland in the category of minor middle powers. We will return to the question of the classification of Poland in our analysis of its defence strategies.

Table 2.2 Average military expenditures 2010–2019.

	Military expenditures in USD millions (world rank)	Military expenditures as percentage of GDP
Poland	10 004 (23)	1.90
Romania	2 986 (51)	1.46
Czechia	2 279 (56)	1.06
Hungary	1 421 (64)	1.01
Slovakia	1 138 (66)	1.17
Croatia	971 (71)	1.69
Bulgaria	898 (72)	1.53

Analysing and explaining strategic adjustment and diversity 27

	Military expenditures in USD millions (world rank)	Military expenditures as percentage of GDP
Lithuania	584 (79)	1.35
Slovenia	539 (80)	1.11
Estonia	491 (86)	1.92
Latvia	399 (95)	1.30

Source: SIPRI (2020).

Regarding diplomatic recognition measured as hosting foreign embassies, Poland has a narrow lead over Hungary, Czechia and Rumania (see Table 2.3). Notably, Slovakia does not reach the top six regarding diplomatic representation. At the same time, Bulgaria's relatively high score on this indicator is considered not being enough to qualify as a minor middle power. Consequently, only Czechia, Hungary, Poland and Romania qualify for the positive 70–70–70 criterion and are hence considered to be minor middle powers. Bulgaria, Croatia, Estonia, Latvia, Lithuania, Slovakia and Slovenia are categorized as being small states. However, based on face validity, diplomatic recognition seems to be the weakest of our indicators on latent power resources. Since Slovakia, although narrowly, met our criteria for economic and political power resources, we will evaluate the classification of Slovakia in our analysis of its defence strategies to see whether or not its strategies correspond to the four minor middle powers.

Table 2.3 Foreign embassies in the 11 new European allies.

	Hosted foreign embassies
Poland	96
Czechia	85
Hungary	84
Romania	84
Bulgaria	71
Croatia	56
Slovakia	44
Slovenia	36
Lithuania	37
Latvia	37
Estonia	33

Source: Diplomatic list presented by each of the 11 states during 2018–2021.

Regarding how differences in relative power affect states' defence strategies, there is no previous research on minor middle powers specifically. Therefore, we have to base our assumptions regarding expected outcomes on previous research on middle powers and small states. If differences in relative power among the 11 new allies correspond to differences in the strategies, we should expect that the minor middle powers share some similarities with other middle powers that the six states categorised as small states do not.

Researchers on the alignment and military strategies of small states have argued that power asymmetries between this category of states and more resourceful states make small states more sensitive to changes in the balance of power and more dependent on support from other states or institutions. According to an influential definition by Robert Rothstein, a 'small power is a state which recognises that it cannot obtain security primarily by use of its own capabilities and that it must rely fundamentally on the aid of other states, institutions, processes or developments to do so' (Rothstein 1968:29). Small states' dependency on other actors and institutions forces them to adjust quickly to changes in their external environment, such as the breakdown of systems for collective security, increased tension between great powers and unfavourable changes in the distribution of power between the main competing regional or global great powers. Moreover, small states have greater reasons to see threats to their survival, and a lack of resources forces them to concentrate their resources on short-term and local matters (Rothstein 1968:30–37; Elgström 2000; Edström et al. 2019). According to Annette Baker Fox, small states – in contrast to great powers – often practice 'anti–balance of power' strategies through complying with demands from expansive and threatening great powers (Baker Fox 1959:53).

Rothstein's emphasis on the psychological dimension – that political leaders of small states must be *aware* of their own state's limited capabilities – is an important point since political leaders and governments act on the basis of more or less correct assessments of both the capabilities of their own state and the intentions and capabilities of other states. A small state that pursues a unilateral strategy against a great power and disregards the asymmetry in power resources is, according to this definition, not exercising a small state strategy. Rothstein's definition makes it reasonable to expect that small states should try to compensate for the lack of nationally controlled military resources by using external efforts (alignment strategies). Great powers and major middle powers, on the other hand, can rely more on internal efforts (military strategies) and on their own power resources to promote their interests. Our previous study on small states confirmed the general expectations regarding these states' awareness of the limitations of the state's own military capacity, their perceived dependence on other actors and institutions, and the corresponding preference for multilateral cooperation.

However, previous research on middle powers also emphasises their preferences for multilateral cooperation. Researchers focusing on common behavioural patterns of statecraft among middle powers have identified common elements such as coalition building using entrepreneurial and/or technical leadership (Cooper 1997:6 and 9–13). Having access to greater economic, military and political power resources, middle powers have been assumed to have greater interests in global affairs, greater abilities to project power both within and outside their own region, greater access to international decision-making institutions and higher ranking as partners (Holbraad 1984:4–5; Gilley and O'Neil 2014b:4–5). During the Cold War, middle power diplomacy used mediation and bridge-building activities (Henrikson 1997). In the post–Cold War era, middle powers have widened their repertoire of activity on an issue-specific basis using 'niche diplomacy' (Cooper 1997). Nevertheless, middle powers are still assumed to share a general preference for multilateralism, confidence building measures and conflict reduction in promoting international security (Nossal and Stubbs 1997:149–151). Following this, we would expect that middle powers perceive themselves to be more dependent on external efforts than great powers but less dependent than small powers. We would also expect that middle powers have greater ambitions regarding their internal efforts than small states but lower ambitions in comparison to great powers. Regarding military strategy, we expect that minor middle powers have higher ambitions compared to small states when it comes to goals related to influence and status and that they have higher ambitions related to capacities regarding unilateral or collective national defence and contributions to expeditionary warfare.

2.2.2 Influence of geography and historical experiences

Both classical realist scholars and researchers within the field of Strategic Studies have acknowledged the importance of geography. As an example of the former, Morgenthau argued that geography is 'the most stable factor upon which the power of a nation depends' (Morgenthau 2006:122–124). In a similar manner, arguing in support of the recommendation that the US should pursue a policy of 'restraint' and pursue a strategy of 'off-shore balancing', contemporary realists such as Christopher Layne (2012) and John Mearsheimer (2018) partly base their arguments on the US's uniquely protected position as the single great power and regional hegemon in its own geographic region. A leading scholar in the field of Strategic Studies, Colin Gray, has argued that 'as a limitation upon the power of states, nothing has proven to be more pervasive and enduring than geography', and it is 'close to self-evident' that 'geographical factors (location, size and character of national territory, character of neighbours and so forth) must permeate defensive thinking'. Moreover, physical and political geography provide opportunities and dangers that 'conditions the frame of reference for the official and public debate over national choices in policy and grand strategy'. Admittedly, the geographical circumstances, Gray continued, do not determine the details of grand strategy, but they shape 'the policy and strategic problems in need of solution'. According to Gray, geographical characteristics also contribute to a nation's strategic culture. In tandem with culture, geography 'predisposes states and their military establishments towards particular ways in warfare'. At the same time, 'the objective aspects of geography [. . .] can have a logic of their own' (Gray 2006:137–139 and 146).

Due to our great number of cases, this variable will not be fully analysed. Instead, we focus on one central geographical aspect, a shared land border with Russia. Arguably, the change in the regional security environment that occurred after Russia's armed aggression against Georgia and Ukraine, respectively, is most alarming to states bordering Russia. Therefore, we are curious to see if this single geographical characteristic is able to produce similarities between states that have a different position in the international system and have different historical experiences. If this variable proves to be important, we expect the four states sharing a land border with Russia (Estonia, Latvia, Lithuania and Poland) to be more reluctant to follow the post-national trends discussed in the previous section and to dismantle military resources related to national defence, especially land forces. Moreover, we expect these four states to prioritise national defence to a greater extent than states without land borders with Russia and to respond more firmly to the deteriorating regional security order following Russia's armed attacks against its neighbours.

However, differences in relative power may create differences within this subcategory as well. Poland, having access to a much greater latent power resources may have opportunities for using its own military means that the three Baltic small states do not have and hence to create costs for military aggression, even against a great power such as Russia. Additionally, Poland has a greater strategic depth and can more easily receive military support from NATO allies through bordering allies in the west and the south. Estonia, Latvia and Lithuania have all very little strategic depth and are therefore assumed to be more dependent on external support and preparations for collective self-defence to create thresholds for armed aggression by Russia.

The influence of the *previous experiences of armed conflicts* is well documented. This influence may take the form of 'historical lessons', in which case historical examples may either be a part of a learning process or used as a way to legitimise a particular course of action (Snyder 1991; Levy 1994). According to Jack Levy, '[P]eople learn more from failure than from success' and 'past success contributes to policy continuity whereas failure leads to policy change' (Levy 1994:304). War experiences are particularly important formative experiences. Experiences received during 'systematic wars' between great powers have, according to Dan Reiter, the greatest impact on future choices between alternative strategies (Reiter 1994). Historical experiences relating to both wars and other formative experiences during state-building processes have also been analysed as one of several sources of national strategic cultures. Darryl Howlett and Jeffrey Lantis have argued that national conceptions of roles and identities may be characterised as 'axiomatic beliefs', a collective consciousness regarding a nation state's relation to its external environment as well as its understanding of appropriate ways of acting. This makes these intersubjective understandings relatively resilient to change as they serve as a filter for future learning (Lantis and Howlett 2016:89–94).

However, changes in both identity and culture do occur. External shocks that challenge existing beliefs and undermine past narratives is one potential source of change. Defeat in war is an example of an external shock with a great potential to produce sudden and dramatic changes in previously axiomatic belief systems. Regardless of whether historical experiences are a part of a learning process or are simply used as a rhetorical tool, they affect the freedom of action for policymakers. As Stanley Hoffmann once argued: in advocating a specific foreign policy move, the policymaker 'can do so only by taking the nation along, with its baggage of memories and problems' (Hoffmann 1995:75–76). This makes some policy options (those associated with previous failures or those not in line with national traditions) more difficult to promote, while other policy options may instead find support in the collective intersubjective interpretations of the previous experiences of a particular national community.

According to Hoffmann, the diversity of national experiences and self-conceptions create a 'logic of diversity', which complicates agreements on common positions and strategies among states involved in the European integration project. Adrian Hyde-Price argues that the diversity of strategic cultures risks creating obstacles to efforts of establishing 'a common European and/or transatlantic security policy relevant to the post–Cold War security environment'. The question that many European countries now face is, according to Hyde-Price, whether a strategic culture largely formed in the mid-twentieth century is still appropriate. Strategic cultures tend to 'outlast the era of their original inception', resulting in 'embedded strategic cultures' that can become an impediment to the conceptual and policy change needed to formulate responses to a new and changing security environment (Hyde-Price 2004:326–327).

In our previous study on small states, we noticed that historical experiences relating to WWI and WWII had a significant and lasting impact on the strategic priorities of all four Nordic countries. Based on how the Nordic countries' different war experiences affected their defence strategies during both the Cold War and post–Cold War era, four dimensions of historical experiences seem to be of particular importance (Edström *et al.* 2019). The *first* dimension concerns involvement or non-involvement in great power wars, the *second* experiences of national defence against aggression and hence especially from great powers, the *third* experiences of offensive warfare in alliance with Nazi-Germany and/or the USSR, and the *fourth* experiences of military assistance (MA) from and/or cooperation with one or more Western great powers. In analysing the influence on experiences during the systematic wars, we will restrict our attention to WWII.

Since all 11 new allies, one way or another, were involved in WWII, the first dimension is not of any further interest for this study. However, based on the experiences from WWII, which are given specific attention in our analysis, we claim that two main categories of states can be identified. The first group consist of the states that suffered not only from intrusion by one or several major powers but also from annexation and ceased to exist as independent states. The members of this category are Czechia, Estonia, Latvia, Lithuania, Poland and Slovenia. The other category consists of the states that were able to keep at least some form of independence during WWII but despite all efforts suffered from occupation. While Croatia and Slovakia were established as German puppet states, Bulgaria, Hungary and Romania kept their independence. Notably, all five states in this category fought with Nazi-Germany in the initial phases of WWII and were occupied by the USSR during the final phase. Regarding the experiences of national defence, we expect that states having positive experiences, despite suffering an armed aggression from another state, will be more reluctant to dismantle their capacities for national defence and transform their armed forces for expeditionary warfare. States that have less or no positive experiences of national defence (including surrendering without resistance) after suffering an armed aggression will give greater priority to efforts related to collective defence, especially if they consider themselves threatened by a more resourceful state.

Experiences of offensive warfare in alliance with a great power, as well as experiences of military assistance from and/or military cooperation (MC) with great powers, concern relations to specific great powers. Are states that were invaded and occupied by the USSR responding more harshly to a renewed potential Russian military threat compared to states that were not fighting the USSR? Do states that suffered attacks from Nazi-Germany find it more difficult to cooperate with the EU and/or NATO in matters related to collective defence? Are states with positive experiences from joining forces with great powers more inclined to commit themselves to allied war efforts compared to states with negative experiences? Contrary to Hyde-Price, we consider other formative periods of state-building, both before and after the WWII–Cold War era, potentially being relevant in this regard as well. Clearly, the 11 new allies have different experiences from the middle ages. Bulgaria, Croatia, Hungary, Lithuania, Poland, Poland/Lithuania were all independent states, although for different lengths of time. Czechia and Romania can both track their roots back to being semi-independent states – to Bohemia in the former case and to Moldavia, Transylvania and Wallachia in the latter case. Four countries, Estonia, Latvia, Slovakia and Slovenia, lack the experience of being independent during the middle ages. Also in the more recent perspective, the 11 new allies have different experiences. While four states gained independence prior to WWII, that is Romania in 1878, Bulgaria in 1878/1908, as well as Hungary and Poland in 1918, the seven other states gained their *lasting* independence only after the end of the Cold War. Despite the challenges involved, we will briefly address this aspect of different historical experience as well. Obviously, one such challenge is the fact that some of the seven latter countries, that is the three Baltic States, as well as Czechia's predecessor Czechoslovakia, were temporarily independent during the interwar era. Moreover, the way independence after the Cold War was achieved differs. Croatia and Slovenia had to fight a civil war; Estonia, Latvia and Lithuania had to struggle with the Soviet occupation forces, while Czechoslovakia was peacefully divided into Czechia and Slovakia. Presumably, states that gained lasting independence only after the end of the Cold War will give priority to ends related to protecting this new or regained independence, while the former group rather will prioritise influence or improving their position in the international system. This may also affect the views on the means, with the latter category focusing on national defence and the former on using their military expeditionary in order to gain status and/or influence.

We will return to these historical events in the initial section in each of the country-specific chapters. For now, we consider that the main dividing line regarding historical experience is experience from WWII. As a potential reinforcing factor, we will also include the time of lasting independence prior to WWII versus after the Cold War in our analysis.

2.2.3 Strategic exposure

Together, the three variables just elaborated on will affect the extent to which states perceive themselves as being *strategically exposed*. Different levels of perceived strategic exposure is expected to affect how governments respond to external pressures to develop capacities for expeditionary warfare or common efforts related to collective self-defence. Small states have greater reasons to feel threatened by more resourceful states, and they may therefore be more reluctant to dismantle their national defences. On the other hand, if the power asymmetry between the less resourceful state and the perceived threatening state is too great, the small state will have to rely mainly on external efforts, practicing alignment strategies by seeking support from friendly powers or appeasing the threatening state. Additionally, relative power may affect the level of ambition when it comes to expeditionary warfare, whereas the lack of resources makes it difficult for small states to give substantive contributions to common efforts. Obviously, a land border to a great power that is perceived as threatening will increase the perceived strategic exposure, making the less resourceful state even more reluctant to dismantle capacities for national defences and more dependent on developing alignment strategies for external support or appeasing the threatening power. Historical experiences are likely to affect both the level of ambition when it comes to national defence and expeditionary warfare and the perceptions related to specific states and organisations when it comes to identifying preferred institutional platforms, trustworthy partners and friendly and threatening powers.

In Table 2.4, the potential aggregated perception of strategic exposure, based on the three intervening variables, are summarised. The 11 new allies' different experiences during WWII are not included in the table, but we will return to this issue in the empirical chapters and in the concluding chapters. Clearly, we expect the three Baltic States to present the highest degree of perceived exposure and Hungary and Romania to present the lowest. In the third and last part of this

Table 2.4 Potential aggregated perception of strategic exposure.

	Relative power Small state?	Geography Sharing land border with Russia?	Historical experience Independence after Cold War?
Bulgaria	Yes		
Croatia	Yes		Yes
Czechia			(heir of Czechoslovakia)
Estonia	Yes	Yes	Yes
Hungary			
Latvia	Yes	Yes	Yes
Lithuania	Yes	Yes	Yes
Poland		Yes	
Romania			
Slovakia	Yes		Yes
Slovenia	Yes		Yes

book, we will evaluate to which extent our expectations on covariance between the intervening and dependent variables have been met.

Notes

1 For an analysis of similarities and differences among the Nordic states' defence transformation processes, see Jakobsen (2005), Matlary and Østerud (2007), Edström *et al.* (2019) and Christiansson (2020). In Edström and Westberg (2020a), as well as in Edström and Westberg (2021), similar developments among middle powers and great powers situated in regions characterized by low war expectations are described.
2 For a discussion on the possibility to combine assumptions associated with realism and research on strategic culture, see Rynning (2011).
3 See, for example, Rose (1998), Schweller (2004), Rathbun (2008).
4 See, for example, Baker Fox (1959), Vital (1967), Rothstein (1968), Holbraad (1984), Wight (1986), Gilley and O'Neil (2014a).
5 A similar hierarchical conception of the state system is described by Øyvind Østerud who uses four categories (1) superpowers and great powers, (2) middle powers, (3) small powers and (4) micro-states (Østerud 1992).
6 On the contested nature of the middle power concept, see Chapnick (1999), Jordaan (2003), Ping (2005), Ungerer (2007), Cooper (2011), Gilley and O'Neil (2014b), Manicom and Reeves (2014), Patience (2014). For a survey of different trends in small state research and various approaches to defining small states, see Neumann and Gstöhl (2006) and Edström *et al.* (2019).

Bibliography

Baker Fox, Annette (1959). *The Power of Small States: Diplomacy in World War II*. Chicago: University of Chicago Press.
Brezhnev, Leonid (2006). 'The Brezhnev Doctrine' in Gearóid Tuathail, Simon Dalby and Paul Routledge (eds). *The Geopolitics Reader*. Abingdon: Routledge.
Brooks, Stephen and William Wohlforth (2016a). 'The Rise and Fall of Great Powers in the Twenty-First Century: China's Rise and the Fate of America's Global Position' *International Security* Volume 40, Issue 3.
―――― (2016b). *America Abroad: The United States' Global Role in the 21st Century*. Oxford: Oxford University Press.
Bulgarian Ministry of Foreign Affairs (2021). *Diplomatic Corps List*. (www.mfa.bg/en/topical-information/protocol-guide/diplomatic-corps-list).
Bush, George W. (2001). 'Remarks by the President in Address to Faculty and Students of Warsaw University' 15 January, Warsaw University. (https://georgewbush-whitehouse.archives.gov/news/releases/2001/06/text/20010615-1.html).
Buzan, Barry and Ole Wæver (2003). *Regions and Powers: The Structure of International Security*. Cambridge: Cambridge University Press.
Carr, Andrew (2014). 'Is Australia a Middle Power? A Systematic Impact Approach' *Australian Journal of International Affairs* Volume 68, Issue 1.
Chapnick, Adam (1999). 'The Middle Power' *Canadian Foreign Policy Journal* Volume 7, Issue 2.
Christiansson, Magnus (2020). *Defence Transformation in Sweden: The Strategic Governance of Pivoting Projects 2000–2010*. Stockholm: Santérus Academic Press.
Cooper, Andrew (1997). 'Niche Diplomacy: A Conceptual Overview' in Andrew Cooper (ed). *Niche Diplomacy: Middle Powers after the Cold War*. London: Macmillan Press.

Cooper, Andrew, Richard Higgott and Kim Nossal (1993). *Relocating Middle Powers: Australia and Canada in a Changing World Order*. Vancouver: University of British Colombia Press.

Cooper, David (2011). 'Challenging Contemporary Notions of Middle Power Influence: Implications of the Proliferation Security Initiative for "Middle Power Theory"' *Foreign Policy Analysis* Volume 7, Issue 3.

Croatian Ministry of Foreign and European Affairs (2021). *Diplomatic Missions and Consular Offices to Croatia*. (www.mvep.hr/en/diplomatic-directory/diplomatic-missions-and-consular-offices-to-croatia/).

Czech Ministry of Foreign Affairs (2021). *Diplomatic List*. (www.mzv.cz/file/442309/Diplomatic_List).

de Flers, Nicole Alecu and Patrick Müller (2012). 'Dimensions and Mechanisms of the Europeanization of Member State Foreign Policy: State of the Art and New Research Avenues' *Journal of European Integration* Volume 34, Issue 1.

Diesen, Sverre (2005). 'Mot et alliansintegrert forsvar' in Øyvind Østerud and Janne Haaland Matlary (eds). *Mot et avnasjonalisert forsvar?*. Oslo: Abstrakt forlag.

Drent, Margriet, David Greenwood, Sander Huisman and Peter Volten (2001). *Organising National Defences for NATO Membership: A Report*. Groningen: Centre for European Security Studies.

Drulàk, Petr (2012). 'Czech Geopolitics: Struggling for Survival' in Stefano Guzzini (ed). *The Return of Geopolitics in Europe? Social Mechanisms and Foreign Policy Identity Crises*. Cambridge: Cambridge University Press.

Edmunds, Timothy (2005). 'A New European Security Environment? The Evolution of Military Roles in Post-Cold War Europe' in Timothy Edmunds and Marjan Malesic (eds). *Defence Transformation in Europe: Evolving Military Roles*. Amsterdam: IOS Press.

Edström, Håkan, Dennis Gyllensporre and Jacob Westberg (2019). *Military Strategy of Small States: Responding to the External Shocks of the 21st Century*. Abingdon: Routledge.

Edström, Håkan and Jacob Westberg (2020a). *Military Strategy of Middle Powers: Competing for Security, Influence and Status in the 21st Century*. Abingdon: Routledge.

――― (2020b). 'Between the Eagle and the Bear: Explaining the Alignment Strategies of the Nordic Countries in the 21st Century', *Comparative Strategy* Volume 39, Issue 2.

――― (2021). *Military Strategy of Great Powers: Managing Power Asymmetries & Structural Changes in the 21st Century*. Abingdon: Routledge.

Elgström, Ole. (2000). *Images and Strategies for Autonomy: Explaining Swedish Security Policy Strategies in the 19th Century*. London: Kluwer Academic Publishers.

Estonian Ministry of Foreign Affairs (2018). *The Tallinn Diplomatic List*. (https://vm.ee/sites/default/files/content-editors/state-protocol/diplomatic_list_2018_51.pdf).

European Council (1993). *Conclusion of the Presidency*. European Council in Copenhagen 21–22 June. (www.consilium.europa.eu/media/21225/72921.pdf).

Fukuyama, Francis (1992). *The End of History and the Last Man*. London: Hamish Hamilton.

Gilley, Bruce and Andrew O'Neil (eds) (2014a). *Middle Powers and the Rise of China*. Washington, DC: Georgetown University Press.

――― (2014b). 'China's Rise through the Prism of Middle Powers' in Bruce Gilley and Andrew O'Neil (eds). *Middle Powers and the Rise of China*. Washington, DC: Georgetown University Press.

Gray, Colin (1999). 'Strategic Culture and Context: The First Generation Strikes Back' *Review of International Studies* Volume 25, Issue 1.

—— (2006). *Strategy and History: Essays on Theory and Practice*. Abingdon: Routledge.

Haushofer, Karl (2006) 'Why Geopolitics?' in Gearóid Tuathail, Simon Dalby and Paul Routledge (eds). *The Geopolitics Reader*. Abingdon: Routledge.

Henrikson, Alan. (1997). 'Middle Powers as Managers: International Mediation within, across and Outside Institutions' in Andrew Cooper (ed). *Niche Diplomacy: Middle Powers after the Cold War*. London: Macmillan Press.

Hoffmann, Stanley (1995). *The European Sisyphus: Essays on Europe 1964–1994*. Boulder, CO: Westview Press.

Holbraad, Carsten (1971). 'The Role of Middle Powers' *Cooperation and Conflict* Volume 6, Issue 1.

—— (1984) *Middle Powers in International Politics*. London: Macmillan Press.

Hungarian Ministry of Foreign Affairs and Trade (2018). *Diplomatic Corps*. (https://vm.ee/sites/default/files/content-editors/state-protocol/diplomatic_list_2018_51.pdf).

Huntington, Samuel (1993). 'The Clash of Civilizations?' *Foreign Affairs* Volume 72, Issue 3.

Hyde-Price, Adrian (2004) 'European Security, Strategic Culture, and the Use of Force' *European Security* Volume 13, Issue 4.

IMF (2021). *World Economic Outlook Database*. (www.imf.org/en/Publications/WEO/weo-database/2021/April).

Jakobsen, Viggo (2005) 'Stealing the Show: Peace Operations and Danish Defence Transformation' in Timothy Edmunds and Marjan Malesic (eds). *Defence Transformation in Europe: Evolving Military Roles*. Amsterdam: IOS Press.

Jordaan, Eduard (2003). 'The Concept of Middle Power in International Relations: Distinguishing between Emerging and Traditional Middle Powers' *Politikon* Volume 30, Issue 2.

Kier, Elizabeth (1997). *Imagining War: French and British Military Doctrine between the Wars*. Princeton, NJ: Princeton University Press.

Kuus, Marje (2012). 'Banal Huntingtonianism: Civilizational Geopolitics in Estonia' in Stefano Guzzini (ed.) *The Return of Geopolitics in Europe? Social Mechanisms and Foreign Policy Identity Crises*. Cambridge: Cambridge University Press.

Lantis, John and Darryl Howlett (2016). 'Strategic Culture' in John Baylis, James Wirtz and Colin Gray (eds). *Strategy in the Contemporary World*. Oxford: Oxford University Press.

Lašas, Ainius. (2010). *European Union and NATO Expansion: Central and Eastern Europe*. New York: Palgrave Macmillan.

Latvian Ministry of Foreign Affairs (2021). *Diplomatic List*. (www.mfa.gov.lv/en/about-the-ministry/state-protocol/diplomatic-list).

Layne, Christopher (2012). 'This Time It's Real: The End of Unipolarity and the "Pax Americana"' *International Studies Quarterly* Volume 56, Issue 1.

Legro, Jeffrey and Andrew Moravcsik (1999). 'Is Anybody Still a Realist?' *International Security* Volume 24, Issue 2.

Levy, Jack (1994). 'Learning and Foreign Policy: Sweeping a Conceptual Minefield' *International Organization* Volume 48, Issue 2.

Lieven, Anatol and Dmitrij Trenin (ed.) (2003). *Ambivalent Neighbors: The EU, NATO, and the Price of Membership*. Washington, DC: Carnegie Endowment for International Peace.

Lithuanian Ministry of Foreign Affairs (2021). *Foreign Representations and Honorary Consuls*. (www.urm.lt/default/en/embasycontacts/listing/category.2).

Mackinder, Halford (2006). 'The Geopolitical Pivot of History' in Gearóid Tuathail, Simon Dalby and Paul Routledge (eds). *The Geopolitics Reader*. Abingdon: Routledge.

Manicom, James and Jeffrey Reeves (2014). 'Locating Middle Powers in International Relation Theory' in Bruce Gilley and Andrew O'Neil (eds). *Middle Powers and the Rise of China*. Washington, DC: Georgetown University Press.

Matlary, Janne Haaland (2009). *European Union Security Dynamics: In the New National Interest*. Basingstoke: Palgrave Macmillan.

Matlary, Janne Haaland and Øyvind Østerud (2007). *Denationalisation of Defence: Convergence and Diversity*. Aldershot: Ashgate.

Mearsheimer, John (2018). *The Great Delusion: Liberal Dreams and International Realities*. New Haven, CT: Yale University Press.

Morgenthau, Hans (2006). *Politics among Nations: The Struggle for Power and Peace*. New York: McGraw-Hill.

Moumoutzis, Kyriakos and Sotirios Zartaloudis (2016) 'Europeanization Mechanisms and Process Tracing: A Template for Empirical Research' *Journal of Common Market Studies* Volume 54, Issue 2.

Neumann, Iver and Sieglinde Gstöhl (2006). 'Lilliputians in Gulliver's World?' in Christine Ingebritsen, Iver Neumann, Sieglinde Gstöhl and Jessica Beyer (eds). *Small States in International Relations*. Reykjavik: University of Iceland Press.

Nossal, Kim Richard and Richard Stubbs (1997). 'Mahathir's Malaysia: An Emerging Middle Power?' in Andrew Cooper (ed). *Niche Diplomacy – Middle Powers after the Cold War*. London: Macmillan.

Østerud, Øyvind (1992). 'Regional Great Powers' in Iver Neumann (ed). *Regional Great Powers in International Politics*. London: Macmillan.

Patience, Allan. (2014). 'Imagining Middle Powers' *Australian Journal of International Affairs* Volume 68, Issue 2.

Ping, Jonathan (2005). *Middle Power Statecraft: Indonesia, Malaysia and the Asia-Pacific*. Aldershot: Ashgate.

Polish Ministry of Foreign Affairs (2021). *Diplomatic Missions in Poland*. (file:///C:/Users/iss11065/Downloads/LISTA_CD_Misje_dyplomatyczne_30_06_2021.pdf).

Rasmussen, Mikkel Vedby (2006). *The Risk Society at War: Terror, Technology and Strategy in the Twenty-First Century*. Cambridge: Cambridge University Press.

Rathbun, Brian (2008). 'A Rose by Any Other Name: Neoclassical Realism as the Logical and Necessary Extension of Structural Realism' *Security Studies* Volume 17, Issue 2.

Reiter, Dan (1994). 'Learning, Realism, and Alliances: The Weight of the Shadow of the Past' *World Politics* Volume 46, Issue 4.

Robertson, Jeffrey (2017). 'Middle-Power Definitions: Confusion Reigns Supreme' *Australian Journal of International Affairs* Volume 71, Issue 4.

Romanian Ministry of Foreign Affairs (2021). *Foreign Missions*. (www.mae.ro/en/foreign-missions).

Rose, Gideon (1998). 'Neoclassical Realism and Theories of Foreign Policy' *World Politics* Volume 51, Issue 1.

Rothstein, Robert (1968). *Alliances and Small Powers*. New York: Columbia University Press.

Rynning, Sten (2011). 'Strategic Culture and the Common Security and Defence Policy: A Classic Realist Assessment and Critique' *Contemporary Security Policy* Volume 32, Issue 3.

Schweller, Randall. (2004). 'Unanswered Threats: A Neoclassical Realist Theory of Underbalancing' *International Security* Volume 29, Issue 2.

Simon, Jeffrey (2001). 'NATO's Membership Action Plan and Defense Planning' *Problems of Post-Communism* May/June.

SIPRI (2020). *Military Expenditure Database*. (www.sipri.org/databases/milex).

Slovakian Ministry of Foreign and European Affairs (2021). *The Diplomatic Corps List*. (www.mzv.sk/web/en/the-diplomatic-corps-list).

Slovenian Ministry of Foreign Affairs (2021). *Diplomatic List*. (www.gov.si/en/state-authorities/ministries/ministry-of-foreign-affairs/about-the-ministry/directorate-for-international-law-and-protection-of-interests/diplomatic-protocol/).

Smith, Rupert (2007). *The Utility of Force: The Art of Warfare in the Modern World*. New York: Alfred a Knopf.

Snyder, Jack (1991). *Myths of Empire: Domestic Politics and International Ambition*. Ithaca, NY: Cornell University Press.

Truman, Harry (2006). 'The Truman Doctrine' in Gearóid Tuathail, Simon Dalby and Paul Routledge (eds). *The Geopolitics Reader*. Abingdon: Routledge.

Tuathail, Gearóid (2006). 'General Introduction: Thinking Critical about Geopolitics' in Gearóid Tuathail, Simon Dalby and Paul Routledge (eds). *The Geopolitics Reader*. Abingdon: Routledge.

Ungerer, Carl. (2007). 'The "Middle Power" Concept in Australian Foreign Policy' *Australian Journal of Politics and History* Volume 53, Issue 4.

Vital, David (1967). *The Inequality of States: A Study of the Small Power in International Relations*. Oxford: Calderon Press.

Waltz, Kenneth (1979). *Theory of International Politics*. Long Grove, IL: Waveland Press.

——— (1993). 'The Emerging Structure of International Politics' *International Security* Volume 16, Issue 3.

Wight, Martin (1986). *Power Politics*. Harmondsworth, UK: Pelican Books/RUSI.

3 Operationalising the dependent variable

Defence strategy

3.1 Introduction

In his seminal work *Theory of International Politics*, Kenneth Waltz argued that each state must have its own the survival as its most fundamental end. In a self-help system, the means for self-preservation fall into two categories: (1) internal efforts, that is measures to increase the state's military strength and resilience, and (2) external efforts, that is measures to strengthen the state's own military alliance and/or weaken that of opposing state (Waltz 1979). Our definition of defence strategy combines these two categories. *Defence strategy* is defined as interconnected ideas on how politically defined strategic ends should be achieved through a combination of alignment strategies and suitable strategic ways of developing and employing military means. *Alignment strategies* refer to different ways of interacting on a political level with other states and organizations to promote the state's own interests relating to security, influence or status. This aspect of strategy is part of a states' external efforts to promote its perceived interest. Examples of alignment strategies are balance of power, bandwagoning, isolation and hedging. *Military strategy* concerns the creation, direction and use of military force. This aspect of strategy focuses on state's internal efforts to promote its interest by developing and using the state's own military resources (Edström *et al.* 2019). More specifically, we will approach the concept of military strategy through the lens of Maxwell Taylor's definition that frames strategy as a matching set of ends, ways and means (Lykke 1989). Diplomatic and economic strategies, which do not concern questions related to military power, are not included in our definition of defence strategy. The defence strategies analysed in this book are therefore less inclusive than most definitions of grand strategy (or most actual security strategies of states and organizations) but more inclusive than most definitions of military strategy.

In this chapter, we define and describe the constitutive elements of our definition of defence strategy. In the first section, we present four basic options for alignment strategies. In the second section, we present our definition and operationalisation of military strategy.

DOI: 10.4324/9781003298052-4

3.2 Alignment strategy

Alignment strategies may be divided into four basic options: (1) balance of power, (2) bandwagoning, (3) isolation, and (4) hedging. These alignment strategies may be pursued both within and outside an alliance, and different members of an alliance may pursue different alignment strategies.[1]

Balance of power strategies are essentially defensive strategies aiming at avoiding losses (Schweller 1994:74). This aim is achieved by creating counterweights to expansive powers in order to increase costs for further expansion. Both first-ranked powers and secondary states are, according to Waltz, expected to 'flock to the weaker side', thereby avoiding their main threat: that one state establish itself as a hegemon (Waltz 1979:126–127). Within an alliance, states can *chain gang*. This strategy includes stronger commitments to agreements on collective defence and offers of military contributions to allied contingency planning or war efforts. Other members may instead *pass the buck* to some or several of their allies, hence freeriding, or rather passive bandwagoning, on the security provided by other members of the alliance. Outside an alliance, states can also pursue a buck-passing strategy or a strategy of *courting*. The latter strategy includes measures to increase the possibility of receiving support from a particular state or a specific alliance, as well as measures to enhance the ability to give and receive military assistance (Christensen and Snyder 1990).

According to Stephen Walt, bandwagoning may be pursued with two distinctly different motives. Defensive bandwagoning may take the form of an *appeasement* policy towards a threatening actor, aiming to avoid an attack by diverting it elsewhere (Walt 1985). Bandwagoning pursued for offensive purposes, 'bandwagoning for profit', involves alignment with the dominant side in a war 'in order to share the spoils of victory' (Schweller 1994:74). For the purposes of this study, *offensive bandwagoning* is defined as strategy primarily motivated by perceived opportunities for gains and includes support to a non-threatening state or alliance. Furthermore, this strategy includes cooperation with the stronger side in a conflict and substantial contributions to common efforts. *Defensive bandwagoning*, on the other hand, is defined as a strategy including unilateral concessions to a threatening state or alliance in order to promote the security of the threatened state. In more recent publications, Walt has updated his analysis of strategic alternatives by focusing on less resourceful states' responses to the US unipolar power (Walt 2005, 2009). The alignment strategy of *regional balancing* presents an additional motive for establishing closer ties with the unipolar power: the desire for protection against a local regional threat. However, since this strategy is primarily directed towards countering a regional threat, it should, in the terminology of balance of threat theory, be considered a specific form of balance of power (Walt 2009). Walt has suggested another modification to the classical balance of power theories, arguing that the choice to either oppose (balance) or ally with (bandwagon) rising powers is based not just on statesmen's assessments of changes in the distribution of power among competing states. The choice of balancing against or allying with

Table 3.1 Alignment strategies.

Balance of power	Bandwagoning	Isolation	Hedging
• Chain-ganging • Courting • Regional balancing	• Offensive (for profit) • Defensive (appeasement) • Passive (buck-passing)	• Active (distancing) • Passive (hiding)	• Leash slipping • Multiple-courting

a rising power is also based on threat perceptions. Thus states will 'ally with or against the most threatening power' (Walt 1985:8–9). The perceived level of threat, Walt argued, is affected not only by changes in the relative distribution of power and factors related to geographic proximity but also by offensive capabilities and perceived intentions. Traditional balance of power theory should therefore be modified into a 'balance of threat theory' (Walt 1987:5 and 21–26). We agree with Walt on this point.

Isolationistic strategies, the third category, may differ in intensity. Active efforts to promote the own state's interests in relation to the great powers is, by Ole Elgström, termed *distancing*. The opposite of this strategy consists of a passive approach, that is avoidance of involvement in great power conflicts – *hiding* (Elgström 2000).

A fourth main alignment strategy is *hedging*. These strategies may include efforts of *multiple-courting*, that is a combination of alignment strategies involving cooperation with several different states or institutional settings. For instance, the EU's policy in relation to Russia has previously sought both to balance a possible aggressive resurgence and to integrate the country in Europe (Art 2004). Ji Yun Lee (2016) describes hedging as a way to avoid the risk of 'betting on the wrong horse'. Walt has identified an additional hedging strategy: *leash slipping*. In applying this strategy, states form an alliance or establish common institutions in order to 'reduce their dependency on the unipole by pooling their own capabilities' (Walt 2009:107). In Table 3.1, the outcomes of the elaboration on alignment strategy are summarised.

3.3 Military strategy

As an academic field of study, Strategic Studies emerged relatively late. However, strategy itself has a long history as a subject studied and taught at military academies with Carl von Clausewitz being a pioneering example and later military thinker and writers such as Basil Liddell Hart and André Beaufre continuing this tradition (Baylis and Wirtz 2007; Larsdotter 2019; Strachan 2019). According to John Baylis and James Wirtz, the scholarly interest in strategy began during the early years of the Cold War era when political leaders and academics from various disciplines wrestled with questions related to survival in the nuclear age. Leading theorists during this era include the physicist Herman Kahn, the Nobel Prize laureate in economics sciences Thomas Schelling, political scientist Bernard Brodie and historian Henry Kissinger. In their view, Strategic Studies cannot be regarded as a discipline in its own right (Baylis and Wirtz 2007). Thomas Mahnken has characterized Strategic Studies as an interdisciplinary field of study exploring issues related to the use of force as an instrument of policy (Mahnken 2003). According to Isabelle Duyvesteyn and James Worrall, the core of the field examines 'the ways in which military power and other coercive instruments may be used to achieve political ends in the course of a dynamic interaction of (at least two) competing wills'. Moreover, Strategic Studies' broad subject matter make it possible to include insights from a variety of academic disciplines, including international relations (IR), political science, sociology and anthropology (Duyvesteyn and Worrall 2017:347).

According to Richard Betts, strategy can be defined as 'the link between military means and political ends, the scheme for how to make one produce the other' (Betts 2000:5). Colin Gray offered a similar definition, claiming that strategy concerns 'the direction and use made of means by chosen ways in order to achieve desired ends' (Gray 2010:18). In addition, Gray made a distinction between *grand strategy* and *military strategy*. The former is defined as 'the direction and use made of any or all the assets' of a state 'including its military instrument, for the purposes of policy as decided by politics'. The latter concerns 'the direction and use made of force and the threat of force, for the purposes of policy as decided by politics' (Gray 2015:47). In his influential work, *The Sources of Military Doctrine*, Barry Posen presented a similar distinction between grand strategy and military doctrine. The latter is defined as 'the subcomponent of grand strategy that deals explicitly with military means' (Posen 1984:13). Our definition of defence strategy is less inclusive than Gray's definition of grand strategy but more inclusive than most definitions of military strategy. Similar to Betts and Gray, we will approach the concept of military strategy through the lens of Maxwell Taylor's definition. It frames strategy as a matching set of ends, ways and means (Lykke 1989). Clausewitz, who coined the ends-means paradigm, initially advanced this view (Herberg-Rothe 2014).

Concerning *ends*, we will analyse the military strategies of our selected cases by focusing on the three basic interests: security, influence and status. The first two basic ends correspond to the two competing views on state interest in the debate between so-called defensive realists (arguing that priority should be given to security and survival) and offensive realists (arguing that priority should be on increasing relative power and influence). One key characteristic of defensive realism is its emphasis on how the structure of the international system, the absence of a central power and the number of main competing powers create pressure on states to adopt similar counterbalancing strategies in order to advance their most fundamental interests. Waltz's argument on this issue rests on two assumptions. The first assumption is that the basic ordering principle of the international system, its anarchical and non-hierarchical nature, forces states to develop self-help strategies. The second assumption is that each state must have its own survival as its most fundamental end since this is a 'prerequisite to achieving any goal that states may have, other than [. . .] promoting their own disappearance as political entities' (Waltz 1979:91–92 and 109–110). In our operationalisation, we use *survival* as one possible prioritized end. Our analysis of the presence or absence of this particular end, we will also include more general considerations related to the protection of the state's political independence and territorial integrity.

The argument that priority should be given to self-preservation and maintaining the state's own relative position in the system has been challenged by offensive realist scholars. Offensive realist scholars argue that states should instead pursue strategies that improve their relative position and power, which increase states' ability to enforce their will on other states (Morgenthau 2006; Mearsheimer 2001, 2018). Governments and people, Morgenthau argued, have many different aims related to fundamental values such as freedom, security and prosperity. However, 'whenever they strive to realize their goal by means of international politics, they do so by striving for power'. Power is therefore always the immediate aim, and increasing the power of the state in relation to other actors is accordingly the key to achieve all other political aims (Morgenthau 2006:29–30). According to John Mearsheimer, great powers do not only strive to increase their relative power. Instead, their 'ultimate aim is to be the hegemon – that is the only great power in the system'. Therefore, great powers are 'rarely content with the current distribution of power' and 'almost always have revisionist intentions' unless the cost and risks of shifting the balance of power are too great (Mearsheimer 2001:2–3). We view increased relative power and influence as a second alternative basic aim. However, since our selection of cases include minor middle powers and small states, our operationalisation of influence includes a broader range of goals related to increasing influence rather than efforts devoted to achieving world hegemony or changing the distribution of power in the regional or global system.[2]

In addition to protecting their own survival or security alongside maximising their influence and ability to enforce their will upon others, states may have strong interests in gaining recognition for having a certain relative positional status rank. *Status* is a relational variable based on mutually recognised differences in status among members of a particular system or organisation. Recognition by other states as having a special status may increase a particular state's political capital, give it a louder voice on certain issues and enable it to have access to an exclusive decision-making forum. In the present state system, recognition includes ideas of each state having equal formal rights to sovereignty, procedures of diplomatic representation and collective recognition by peers, acknowledged as membership and representation in the UN Security Council (UNSC) or in organizations such as the Group of Seven (G7) and the Group of Twenty (G20).

Interest in states' competition for status and recognition is not novel. During the Cold War, some researchers belonging to the so-called English School emphasized this aspect of power competition in their analyses and definitions of great powers and less resourceful states (Bull 1977:200–202; Holbraad 1984:75–76). The insights of the English School regarding status competition were lost in the theoretical debates within the IR discipline in the 1980s and the early 1990s.[3] However, during the latter part of the 1990s and the first two decades of the twenty-first century, interest in status has resurged, resulting in a celebrated edited volume with the title *Status in World Politics*. Deborah Welch Larson and her colleagues define status as 'collective beliefs about a given state's ranking on valued attributes'. Status, they argued, manifests itself in two distinct ways: as membership in a defined club of actors and as a relative standing within a particular club. Membership in 'international society' – recognised sovereignty – is a status sought by many sub-state groups, and once this status is conferred via recognition by others, the new state may continue to improve their relative position further by advancing to middle power status. Ultimately, states may seek entrance to the highest status group – the great power club – and continue to compete for the 'less formal rankings' within this group (Larson *et al.* 2014:7–8).[4] Obviously, the level of ambition regarding status is very different when it comes to small states as compared to those of great powers. However, small states and middle powers may have an interest in gaining recognition from others confirming a certain identity or role, for example, being a competent and trustworthy partner or a respected equal within a particular club (Jakobsen *et al.* 2018; Ångström 2020).

The three basic interests – survival, influence and status – are not mutually exclusive. Power maximisation may be a way to protect the survival of the own state. Increased influence over other actors may increase the status rank of a particular state. Survival may also be taken for granted by states that are situated in a regional context where states do not perceive any existential threats. This, however, does not mean that these 'secure states' do not care about their survival. What matters in our empirical analysis, is what kind of basic aims the political leadership *prioritises* questions related to the use of their armed forces.

When it comes to military *means*, the short-term perspective concerns the quality, quantity and level of preparedness of the armed forces (Collins 2002). In the long-term perspective, the calculus regarding force generation shifts towards the nations' strategic resources, including the projected size of the defence budget, the defence industrial base, manpower, innovation including research and development, military infrastructure as well as the logistical base (Tellis *et al.* 2001). The time between the decision to acquire certain military capabilities and having them operational and fully integrated into the armed forces, tends, however, to be lengthy. The time frames depend on the technology level of the specific state and its ability to convert economic and technological capacities to additional military capabilities (Brooks and Wohlforth 2016). Regarding the use of force, Barry Posen has introduced a useful distinction between means deployed for national defence and those deployed on foreign soil 'based in the theatres in which they would fight' (Posen 2003:18). In a previous study, we introduced a similar distinction between military capabilities primarily developed for *national defence* and capabilities primarily related to *expeditionary warfare* and power projection outside the state's own region (Edström and Westberg 2020a, 2020b). This distinction is useful for this study as well since it focuses on the two main alternative responses to the new and old European security dynamics. Regarding *expeditionary warfare*, we see a spectrum ranging from global power projection via regional power projection to contribution warfare, that is states contributing to common efforts and international military operations led by other states.[5] These different uses of expeditionary warfare reflect differences in both power resources and levels of ambition (Vance 2005). We expect that the new allies' efforts regarding expeditionary warfare will mainly be restricted to contribution warfare.

When addressing the *ways*, that is on how military resources are used or developed, we find few agreements in the literature.[6] In our previous studies on middle powers' strategic adjustment, we focused on one fundamental aspect of how states use and develop their military resources: the choice between *unilateral* or *multilateral* approaches. Since their approach directs the attention to questions related to how far states are prepared to integrate their resources with allies as well as different levels of ambition when it comes to developing unilateral military capabilities, this distinction is applied in this study as well (Edström and Westberg 2020a, 2020b).

To be able to detect and document differences between states regarding priorities of means and ways we have to specify what kind of means characterise each of the two main uses of military force. To be able to contribute to multilateral military operations and military crisis management, states have to develop a number of capacities and capabilities. One such key capability is *interoperability* with armed forces from other states. This includes developing new professional skills and tactics, such as counter-insurgency strategies. Since the opponent in this kind of warfare in most cases is a non-state actor, it creates a need for *smaller, lighter and mobile units* that are able to adjust to a complex and changing conflict environment. Consequently, these capabilities create demands for *professional* soldiers. Regarding military capacities, expeditionary warfare also creates demands on various *logistic chains* and *transport systems* for deployment of military units outside the state's own region.

States prioritising national defence are generally expected to give greater priority to *unilateral* capabilities related to the defence of their own territory. However, due to power asymmetries with potentially threatening states, less resourceful states may also give priority to efforts related to *collective defence* within an alliance. Interoperability with the armed forces of friendly states may therefore also be of importance. Depending on national traditions, states that prioritize national defence are more likely to continue to use *conscription* as a method to mobilize their citizens. If possible, these states will invest in *technologically advanced military systems* such as air defence systems and stealth capacities for warships and airplanes. Notably, these two main uses of military means and ways are not mutually exclusive, and states may try to balance their ambition to be able to pursue both national defence and expeditionary warfare. However, the emphasis may still be on one of the two alternatives.

In Table 3.2, the operationalising of each of the three elements of military strategy are summarised.

Table 3.2 Elements of military strategy.

Ends	Means	Ways
• Survival • Influence • Status	• Capabilities for national defence • Capabilities for international operations (expeditionary warfare)	• Unilateral approaches • Multilateral approaches

Notes

1 The different responses of NATO member states like Denmark and Turkey to a more assertive Russia (that is, prior to 24 February 2022) and the war against the Islamic State (IS) are recent examples of this. When Turkey was approaching Russia with a policy similar to bandwagoning, Denmark was consistently supporting the US in the war against IS and committing itself firmly to NATO's renewed efforts for collective defence in Europe. Regarding Denmark's strategy, see Edström *et al.* (2019).
2 On small states' strategies for gaining influence by contributing to international operations, see, for example, Jakobsen (2009).
3 For a summary of these IR debates, see, for example, Keohane (1986); Hollis and Smith (1990); Jackson and Sørensen (2006).
4 For further discussion on status as a basic interest of states and status competition as an alternative or complementary source for competition and conflicts in IR, see, for example, Schweller (1999), Larson and Shevchenko (2010), Volgy *et al.* (2011), Wohlforth (2011), Thompson (2014), and Renshon (2017).
5 On contribution warfare, see Ångström (2020).
6 For additional interpretations of this element of strategy, see, for example, Buzan (1987), Freedman (1998), Gray (1999), Tangredi (2002).

Bibliography

Ångström, Jan (2020). 'Contribution Warfare: Sweden's Lessons from the War in Afghanistan' *Parameters* Volume 50, Issue 4.
Art, Robert (2004). 'Europe Hedges Its Security Bets' in Paul Thazha and James Wirtz (eds). *Balance of Power Revisited: Theory and Practice in the 21st Century*. Stanford, CA: Stanford University Press.
Baylis, John and James Wirtz (2007). 'Introduction' in John Baylis, James Wirtz, Colin Gray and Eliot Cohen (eds). *Strategy in the Contemporary World: An Introduction to Strategic Studies*. Oxford: Oxford University Press.
Betts, Richard (2000). 'Is Strategy an Illusion?' *International Security* Volume 25, Issue 2.
Brooks, Stephen and William Wohlforth (2016). 'The Rise and Fall of Great Powers in the Twenty-First Century–China's Rise and the Fate of America's Global Position' *International Security* Volume 40, Issue 3.
Bull, Hedley (1977). *The Anarchical Society: A Study of Order in World Politics*. London: Macmillan.
Buzan, Barry (1987). *An Introduction to Strategic Studies*. New York: St. Martin's Press.
Christensen, Thomas and Jack Snyder (1990). 'Chain Gangs and Passed Bucks: Predicting Alliance Patterns in Multipolarity' *International Organization* Volume 44, Issue 2.
Collins, John (2002). *Military Strategy: Principles, Practices, and Historical Perspectives*. Washington, DC: Brassey's.

Duyvesteyn, Isabelle and James Worrall (2017). 'Global Strategic Studies: A Manifesto' *Journal of Strategic Studies* Volume 40, Issue 3.

Edström, Håkan and Dennis Gyllensporre (2013). *Political Aspirations and Perils of Security: Unpacking the Military Strategy of the United Nations*. Basingstoke: Palgrave Macmillan.

Edström, Håkan, Dennis Gyllensporre and Jacob Westberg (2019). *Military Strategy of Small States: Responding to the External Shocks of the 21st Century*. Abingdon: Routledge.

Edström, Håkan and Jacob Westberg (2020a). *Military Strategy of Middle Powers: Competing for Security, Influence and Status in the 21st Century*. Abingdon: Routledge.

——— (2020b). 'Between the Eagle and the Bear: Explaining the Alignment Strategies of the Nordic Countries in the 21st Century' *Comparative Strategy* Volume 39, Issue 2.

Elgström, Ole (2000). *Images and Strategies for Autonomy: Explaining Swedish Security Policy Strategies in the 19th Century*. London: Kluwer Academic Publishers.

Freedman, Lawrence (1998). 'The Revolution in Strategic Affairs' *Adelphi Papers* Volume 38, Issue 318.

Gray, Colin (1999). *Modern Strategy*. Oxford: Oxford University Press.

——— (2010) *The Strategy Bridge: Theory for Practice*. Oxford: Oxford University Press.

——— (2015) *The Future of Strategy*. Cambridge, UK: Polity Press.

Herberg-Rothe, Andreas (2014). 'Clausewitz's Concept of Strategy: Balancing Purpose, Aims and Means' *Journal of Strategic Studies* Volume 37, Issue 6–7.

Holbraad, Carsten (1984). *Middle Powers in International Politics*. London: Macmillan Press.

Hollis, Martin and Steve Smith (1990). *Explaining and Understanding International Relations*. Oxford: Clarendon.

Jackson, Robert and Georg Sørensen (2006). *Introduction to International Relations: Theories and Approaches*. Oxford: Oxford University Press.

Jakobsen, Peter Viggo (2009). 'Small States, Big Influence: The Overlooked Nordic Influence on the Civilian ESDP' *Journal of Common Market Studies* Volume 47, Issue 1.

Jakobsen, Peter Viggo, Jens Ringsmose and Hakon Lunde Saxi (2018) 'Prestige-Seeking Small States' *European Journal of International Security* Volume 3, Issue 2.

Keohane, Robert (ed) (1986). *Neorealism and Its Critics*. New York: Columbia University Press.

Larsdotter, Kersti (2019). 'Military Strategy in the 21st Century' *Journal of Strategic Studies* Volume 42, Issue 2.

Larson, Deborah Welch, T.V. Paul and William Wohlforth (2014). 'Status and World Order' in T.V. Paul, Deborah Welch Larson and William Wohlforth (eds). *Status in World Politics*. New York: Cambridge University Press.

Larson, Deborah Welch and Alexei Shevchenko (2010). 'Status Seekers: Chinese and Russian Responses to US Primacy' *International Security* Volume 34, Issue 4.

Lee, Ji yun (2016). 'Hedging Strategies of the Middle Powers in East Asian Security: The Cases of South Korea and Malaysia' *East Asia* Volume 34, Issue 1.

Lykke, Arthur (1989). 'Toward an Understanding of Military Strategy' in Arthur Lykke (ed). *Military Strategy: Theory and Application*. Carlisle Barracks, PA: US Army War College.

Mahnken, Thomas (2003). 'The Future of Strategic Studies' *Journal of Strategic Studies* Volume 26, Issue 1.

Mearsheimer, John (2001). *The Tragedy of Great Power Politics*. London: WW Norton.

Mearsheimer, John (2018). *The Great Delusion: Liberal Dreams and International Realities*. New Haven, CT: Yale University Press.

Morgenthau, Hans (2006). *Politics among Nations*. New York: McGraw-Hill.
Posen, Barry (1984) *The Source of Military Doctrine: France, Britain, and Germany between the World Wars*. Ithaca, NY: Cornell University Press.
——— (2003). 'Command of the Commons' *International Security* Volume 28, Issue 1.
Renshon, Jonathan (2017). *Fighting for Status: Hierarchy and Conflict in World Politics*. Princeton, NJ: Princeton University Press.
Schweller, Randall (1994). 'Bandwagoning for Profit: Bringing the Revisionist State Back In' *International Security* Volume 19, Issue 1.
——— (1999). 'Realism and the Present Great Power System: Growth and Positional Conflicts Over Scarce Resources' in Ethan Kapstein and Michael Mastanduno (eds). *Unipolar Politics: Realism and State Strategies after the Cold War*. New York: Columbia University Press.
Strachan, Hew (2019). 'Strategy in Theory: Strategy in Practice' *Journal of Strategic Studies* Volume 42, Issue 2.
Tangredi, Sam (2002). 'Assessing New Missions' in Hans Binnendijk (ed). *Transforming America's Military*. Washington, DC: National Defense University Press.
Tellis, Ashley, Janice Bially, Christopher Layne and Melissa McPherson (2001). *Measuring National Power in the Postindustrial Age*. Santa Barbara, CA: Rand Corporation.
Thompson, William (2014). 'Status Conflict, Hierarchies, and Interpretation Dilemmas' in T.V. Paul, Deborah Welch Larson and William Wohlforth (eds). *Status in World Politics*. New York: Cambridge University Press.
Vance, Jonathan (2005). 'Tactics without Strategy, or Why the Canadian Forces Do Not Campaign' in Allan English, Daniel Gosselin, Howard Coombs and Laurence Hickey (eds). *The Operational Art: Canadian Perspectives*. Kingston, ON: Canadian Defence Academy Press.
Volgy, Thomas, Renato Corbetta, Keith Grant and Ryan Baird (eds) (2011). *Major Powers and the Quest for Status in International Politics: Global and Regional Perspectives*. New York: Palgrave Macmillan.
Walt, Stephen (1985). 'Alliance Formation and the Balance of World Power' *International Security* Volume 9, Issue 4.
——— (1987). *The Origins of Alliances*. Ithaca, NY: Cornell University Press.
——— (2005). *Taming American Power: The Global Response to U.S. Primacy*. New York: Norton.
——— (2009). 'Alliances in a Unipolar World' *World Politics* Volume 61, Issue 1.
Waltz, Kenneth (1979). *Theory of International Politics*. Long Grove, IL: Waveland Press.
Wohlforth, William (2011). 'Unipolarity Status Competition and Great Power War' in John Ikenberry, Michael Mastanduno and William Wohlforth (eds). *International Relations Theory and the Consequences of Unipolarity*. Cambridge: Cambridge University Press.

Part II
The empirical exploration

In this part, in order to avoid unintentional clustering, the 11 new European allies are presented in alphabetical order. Moreover, the cases are presented in a similar way. Initially, the primary sources used for the analysis is introduced. Then the historical background is illuminated. This is followed by four sections, analysing both the perceptions regarding the strategic environment and each of the three elements of the military strategy (that is, the ends, means and ways). In the last section, we present our conclusions and relate our results to the findings of previous research.

4 The strategy of Bulgaria[1]

The primary sources analysed in this chapter consist of the defence white papers (DWPs) of 2002 and 2010, the plans for the modernization of the Bulgarian armed forces of 2008, 2010, 2015 and 2017, and the law on defence and armed forces of 2009. Moreover, the national security strategy (NSS) of 2011, the national defence strategy (NDS) of 2011 and the policy paper on Bulgaria's role in the North Atlantic Treaty Organization (NATO) and the European defence of 2015 are studied.[2]

4.1 Historical background

The first Bulgarian Empire was established in 681 and lasted until 1018. During these centuries, the Bulgarians fought several wars against Byzantine, Croatians, Hungarians, Serbians and Kiev-Russians. The Bulgarian Empire was re-established in 1185 but was dissolute in 1396. During these two centuries, the Bulgarians continued to wage war with traditional enemies as well as with Greek states, Wallachia and the Ottomans. Following the Treaty of Berlin in 1878, Bulgaria was recognized as an independent vassal state of the Ottoman Empire. In 1908, Bulgaria declared independence as a sovereign kingdom. From October 1912 to May 1913, Bulgaria, allied with Greece, Montenegro and Serbia, fought the First Balkan War against the Ottomans victoriously. The outcome of the war led to unsolved issues between Bulgaria and its former allies. This led to the Second Balkan War in 1913 in which Bulgaria had to fight not only Greece and Serbia, but Romania as well. The Treaty of Bucharest in August 1913 confirmed Bulgaria's defeat. During the First World War (WWI), Bulgaria aligned with Germany, Austria-Hungary and the Ottomans. This war led to another Bulgarian defeat. In 1941, Bulgaria entered the Second World War (WWII) on the side of Nazi-Germany and participated in the war against Greece and Yugoslavia. Following the changed outcome of the war, Bulgaria switched sides. This attempt was pointless, and in 1944 the country suffered from a Soviet invasion and occupation. In 1946, infant Tsar Simeon II had to resign following the result of a referendum abolishing the monarchy. When the Warsaw Pact (WP) was established in 1955, Bulgaria was one of the founding members (see, for example, Curtis 1993; Liddell-Hart 1997; Crampton 2007).

DOI: 10.4324/9781003298052-6

52 *The empirical exploration*

4.2 Strategic environment

In 2002, the government issued the very first Bulgarian white paper on defence. 'We have tried to reveal the nature of the strategic environment and the context in which the Armed Forces' reform is taking place', Minister of Defence Nikolay Svinarov announced (Bulgarian Council of Ministers (CM) 2002:7). The government observed diminishing risks of global conflict as well as of large-scale military confrontations. The challenges to Bulgaria's security had, according the government, taken new forms such as

> the illegal proliferation of arms and technologies for their production, international terrorism, organised crime as a serious risk factor with social and political implications for the consolidating market economies in South Eastern Europe, ethno-religious and cultural confrontation and extremism, illegal trafficking of people and drugs, biological and chemical threats, ecological risks, natural disasters, major industrial accidents, etc.
> (Bulgarian CM 2002:15)

The regional situations in the Middle East, the Mediterranean, the Caucasus and Central Asia as well as in certain parts of Africa were, due to the presence of strategically important routes and assets, considered as having a wide international impact. The government concluded that 'at present, as well as in the foreseeable future, there is no direct military threat to Bulgaria' (Bulgarian CM 2002:14). Whereas organisations such as the Organization for Security and Co-operation in Europe (OSCE) and the United Nations (UN) were considered as being of importance, the government stressed the increased role of both NATO and the European Union (EU) regarding global crisis management. The United States (US) was perceived having a particularly important role globally, while the importance of Germany, France, the United Kingdom (UK), Italy and Spain mostly was considered in the context of European integration and/or of the stability and security on the European continent. Russia and China were considered as having significant roles in the global security system. Six years later, in 2008, the government proclaimed that Bulgaria was not going to face a threat of a conventional military conflict endangering its sovereignty and territorial integrity during the upcoming decade. If such a threat nevertheless arose, the government promised that the strategic time for warning would allow preparations for the necessary operational capabilities (Bulgarian CM 2008). In 2010, the global financial and economic crisis was in focus. The government concluded that the crisis would affect Bulgaria's defence policy but also that the strategic environment would continue to progress. 'It will be influenced by dynamic and tough-to-predict political, social, technological and military developments. Conflicts will be evermore complex, unpredictable and hard to manage', the government predicted (Bulgarian National Assembly (NA) 2010b:12).

In 2011, the Bulgarian government concluded that while the probability of conventional armed aggression against Bulgaria was minimal, new criminal and transnational risks and threats had arisen. International terrorism, the proliferation of weapons of mass destruction (WMD), regional conflicts, cybercrime and organised crime were hence considered as especially important (Bulgarian NA 2011). Challenges related to climate changes, demographic and ecological problems, the deficiency of energy and natural resources, information security threats, failed states, and the European and Euro-Atlantic integration were also elaborated. 'There is a growing threat of attacking strategic installations on the territory of the Republic of Bulgaria and its population with ballistic missiles from remote countries', the government concluded (Bulgarian CM 2011:7). The government admitted that Bulgaria was exceptionally dependent on one source of energy. 'Therefore, diversification of energy sources and intensified cooperation within the EU is a key necessity for Bulgaria in order to mitigate the negative economic consequences of the crisis to the east of our country', the government concluded (Bulgarian CM 2014:3). Consequently, the development of the ongoing crises in the Middle East and North Africa were considered having the potential of seriously affecting Bulgaria's security and interests. 'Currently, and given the rapidly evolving challenges of modern strategic environment, without NATO Bulgaria does not have the necessary military resources to effectively guarantee its security', the government admitted (Bulgarian CM 2014:7).

4.3 Ends

Defending the national sovereignty and contributing 'towards enhancing regional and European security' were the two fundamental ends of the Bulgarian armed forces, Minister of Defence Nikolay Svinarov declared in 2002 (Bulgarian CM 2002:7). The government stressed that the Bulgarian national values and interests constituted the basic criteria in the formulation and implementation of its policy within the area of security and defence politics. The national interests were clustered into four overarching areas: (1) independence, sovereignty and territorial integrity; (2) democracy, human rights, and the rule of law; (3) sustainable economic development and prosperity; and (4) international peace and security. These objectives were, the government pledged, to be pursued and protected in conformity with universal human values and international law. At the same time, the government explicitly presented its ambition to assert Bulgaria a role as a key political factor in South Eastern Europe. This role was considered crucial in order to establish Bulgaria as an economic centre and for ensuring the full use of the country's scientific and technological potential. All these aspects were not only regarded as ends in themselves but also of primary importance for national security. Clearly, in 2002 joining the West was in focus. 'Bulgaria defines its accession to NATO and the European Union as a strategic goal that corresponds to the long-term interests of the country', the government clarified (Bulgarian CM 2002:20). In 2008, Bulgaria's accession to both NATO and the EU was accomplished. Consequently, the primary goal of the armed forces was said to be defending peace and guaranteeing the country's territorial integrity, sovereignty and security (Bulgarian CM 2008). 'The defence of the Republic of Bulgaria shall', the Assembly

54 *The empirical exploration*

declared in the law on defence and armed forces of 2010, constitute activities 'for providing a stable environment of security and for preparation and realization of military defence of the territory entirety and independence of the state' (Bulgarian NA 2010a, chapter 1, article 3). In the NSS of 2011, several vital national interests were articulated to:

- guarantee the rights, freedom, security and well-being of the individual as well as of the society as a whole;
- defend the sovereignty and territorial integrity of the country as well as the unity of the nation;
- protect the constitutional system and democratic values;
- defend the population and the critical infrastructure;
- promote the national identity;
- ensure the integrity of the Bulgarian civil society;
- secure 'the place that the Republic of Bulgaria deserves to hold in the EU' as well as globally.

(Bulgarian NA 2011:30–31)

Notably, to uphold and promote the national historic values and heritage was declared an additional national interest.

4.4 Means

In 2002, the government stressed the importance of improving interoperability in order to enable Bulgaria's armed forces to cooperate effectively in joint multinational operations under allied command. The organisational structure, composition, training, staffing and equipment were to 'secure the combat readiness and capabilities, adequate to the country's strategic military environment and resource potential' (Bulgarian CM 2002:27). The government gave priority to the resources allocated for participation in international peace support operations. Regarding the land forces, the list included a mechanized brigade, a special forces battalion, a combat engineer battalion, a chemical protection battalion and a nuclear, biological and chemical weapons (NBC) protection company. The naval resources consisted of a corvette, two minesweepers, two auxiliary ships and two ground-based helicopters. The air force was to contribute with four fighters, eight helicopters, three transportation aircraft and an engineer airfield company. The government also announced its ambition to participate in European defence with, amongst other participants, a mechanised battalion, six helicopters and a corvette. In addition, a mechanised brigade, a light infantry battalion, a special forces battalion, some support units, a mixed helicopter squadron, a transportation aviation unit, a frigate, two corvettes, two minesweepers and an auxiliary ship were assigned as deployable forces to be prepared to participate in the NATO forces. It is unclear whether these units were solely predesignated to just one of the three categories or if the very same units were to fulfil tasks in several contexts (Bulgarian CM 2002).

Mobility was a key capability of a strategic dimension regardless of the scale of the deploying forces, the government concluded in 2008 and outlined three complementing approaches: (1) on a multilateral basis within the framework of NATO and the EU, (2) through bilateral agreements, and (3) with the building up of a limited national air and naval potential. The government also provided directions on further developments until 2015. Notably, the numbers of some key equipment remained the same over the period: 160 main battle tanks (MBTs), 378 armoured combat vehicles (ACVs), 192 artillery pieces, 10 transport aircraft, 12 combat helicopters, 18 transport helicopters, 6 combatant ships and six auxiliary ships. However, some key capabilities decreased significantly in numbers. The combat aircraft were, for example, to be reduced from 32 to 20, while all ten combat training aircraft and all six training helicopters were to be taken out of service (Bulgarian CM 2008).

In 2009, the new prime minister, Boyko Borisov, announced his cabinet's ambition establishing 'a clear management programme, in which the implementation of modern defence policy is a priority' (Bulgarian NA 2010b:3). Increasing the expeditionary capacity was hence in focus. In terms of quantity, the ambition regarding the army included either a reinforced battalion or a greater number of smaller units, in total about 1,000 troops. The navy was to participate with resources equivalent to one frigate but only for a period of three to six months per year. The air force was to contribute with helicopters for a period of six months per year (Bulgarian NA 2010b; see also Bulgarian MoD 2010). In 2010, the bulk of the army consisted of two mechanised brigades, one light infantry brigade and one special forces brigade. In 2014, the light infantry brigade was to be transformed to an independent mechanised battalion and the special forces brigade to a regiment. The six ship squadrons organised by the navy was to be reduced to three by 2014. The transformation of the air force mainly focused on the base and command structure (Bulgarian NA 2010b; see also Bulgarian MoD 2010).

Building and developing 'capabilities for participation in missions and operations of NATO, EU, UN, OSCE or other international organizations' were given priority in the NDS issued in 2011 (Bulgarian CM 2011:9). In case of activation of Article 5 of the Washington Treaty, Bulgaria's main contribution was said to be an army brigade tactical group. The government's 'NATO and EU First' approach to defence policy had clear impact on the transformation of the Bulgarian armed forces. 'Bulgaria must make full use of the collective defence industry capabilities of these two organisations. To this end, we shall seek the maximum degree of implementation of projects through the EU and NATO Agencies', the government declared (Bulgarian CM 2014:11). The government listed prioritized capability areas including the interoperability of key fighting equipment, command and control, participating in missions/operations abroad, mobility of the land forces, anti–sea mines operations and self-defence by fighter combat aircraft. Priority was given to 'the acquisition of a new type of core multirole fighter aircraft and the required integrated logistic support. This acquisition will diminish the technological gap with our allies and neighbouring NATO-member countries, as well as our dependence on Russia' (Bulgarian CM 2014:13).

56 *The empirical exploration*

Table 4.1 Main military resources of Bulgaria.

	2000	2010	2020
Army brigade	7	5	2
MBT	1,475	362	90
ACV/APC/IFV	1,994	1,578	280
Combat aircraft	181	62	21
Principle surface combatant	1 frigate, 7 corvettes, 16 missile and patrol craft	4 frigates, 3 patrol craft	4 frigates, 3 patrol craft

Source: International Institute for Strategic Studies (2000, 2010, 2020).

In 2015, the prioritised expeditionary capacity presented in 2010 remained the same with the exception of the air force, now contributing with transport aircraft instead of helicopters. 'Our country will continue to participate with trained military units in EU Battle Groups as a significant element of the effective EU Military Rapid Response Concept. The contribution with military capabilities to potential EU-led operations is performed in line with our contribution to NATO', the government declared (Bulgarian CM 2015:8). By 2020, the bulk of the army was to consist of two mechanised brigades and an independent alpine battalion. The navy was to organise three ship squadrons and to undertake modernisation of the class E-71 frigates. In 2017, Bulgaria joined the EU Permanent Structured Cooperation (PESCO). The government announced its ambition of increasing the defence budget to 2 percent of the gross domestic product (GDP) by 2024. The acquisition of combat aircraft, a multirole patrol vessel and main combat materiel for developing battalion battle groups were hence given priority (Bulgarian CM 2017). See Table 4.1.

4.5 Ways

The government claimed the Bulgarian strategy was 'defensive, responsive and balanced' in nature (Bulgarian CM 2002:45). The use of military force was divided into three separate contexts: (1) contribution to peacetime national security, (2) contribution to global peace and security, and (3) participation in national defence. Regarding the first context, the armed forces were to conduct rescue operations, to protect strategic sites against terrorist attacks and to provide air and sea traffic control and surveillance. When it comes to global peace and security, the armed forces were to take part in multinational military formations conducting conflict prevention as well as other peace support operations. In addition, the armed forces were to conduct humanitarian as well as search and rescue operations both individually and as part of broader international efforts. Finally, regarding national defence, the armed forces were to conduct territorial defence operations in order to defend the country against armed aggression and to take part in NATO collective defence operations outside Bulgarian territory (Bulgarian CM 2002). Some additional operations were added to the list after the Bulgarian memberships in the EU and NATO. The government concluded the most probable employment of the armed forces was in crisis response operations beyond NATO's area of responsibility as well as the territory of the EU. Consequently, this context was given priority, and several types of peace support operations were mentioned including sanctions and embargo enforcement as well as operations against proliferation of WMD. National defensive operation to protect the country's territorial integrity remained central in the elaborations but was now described as part of an allied defensive operation within Bulgaria. Notably, Bulgarian armed forces were not to participate in such operations within the territory of other NATO member states. Bulgarian participation in this context was only discussed related to counterterrorism and antiterrorism operations. 'Multinationality within NATO, the EU or in ad hoc operations will be a common rule for the operations of the Bulgarian Armed Forces', the government announced. 'With the exception of the operations assisting the population and local government in crises of a non-military nature [. . .] all other operations are conducted and will be conducted, to one degree or another, as multinational operations', the government concluded (Bulgarian CM 2008:16).

In 2010, the government defined military activities that were to be realized through 'mutual actions with the NATO allies, with the EU Member States and with international organizations' (Bulgarian NA 2010a, chapter 1, article 6). The threats to Bulgarian national security were said to be met primarily beyond Bulgaria's borders by the participation of Bulgarian expeditionary forces 'in international UN, NATO and EU operations and missions' (Bulgarian NA 2010b:15). Sharing the responsibilities as both a NATO and a EU member 'by effective contribution to collective defense, by participation in conflict prevention and crisis management in non–Article 5 operations and by active engagement in determining and implementing the EU Common Security and Defense Policy' was given priority in the NDS presented in 2011 (Bulgarian CM 2011:10). 'NATO's operation in Afghanistan remains one of the main foreign policy priorities and also the most extended large-scale military commitment for our country', the government declared (Bulgarian CM 2012:30). In addition to the contributions to NATO- and/or EU-led operations, the government also stressed bilateral relations with the US. The fulfilment of the agreement on defence cooperation between the two allies included the build-up of operational joint Bulgarian-US facilities on Bulgarian soil. 'We will continue to develop our strategic partnership with the US in the area of defence', the government proclaimed (Bulgarian CM 2015:10).

4.6 Conclusions: Bulgarian strategy

Arguably, Bulgaria initially adopted a *multiple-courting*, that is hedging, alignment strategy. Right from the beginning, memberships both in NATO and in the EU was a prioritized end. However, Bulgaria not only focused on the two organisations but established bilateral arrangements with the US as well. Due to economic constraints and the increased dependency on the US, we do not exclude that a shift towards a more US-oriented bandwagoning strategy is about to take place. Regarding the ends of the military strategy, we conclude that *survival* is the most appropriate label at the aggregated level. The government has expressed

58 *The empirical exploration*

objectives indicating considerations on status. Concurrently, the necessity of gaining influence both in NATO and in the EU has been explicit in the elaborations. We claim that the continued emphasis on not only the sovereignty and integrity of the Bulgarian state but on the cultural heritage of the Bulgarian nation as well indicates the supremacy of *survival*. However, this end is not explicitly related to an external threat from another state. On the contrary, Bulgaria does not seem to view itself as strategically exposed to threats to its territory from any state despite Russia's war on Georgia and Ukraine, respectively. Consequently, we cannot detect a greater change in Bulgaria's defence strategy following 'the return of the old European security dynamics'. The lack of a land border with Russia appears to be one possible explanation, which we will return to in Part III.

When it comes to means, the Bulgarian government has explicitly declared its priority on *expeditionary* capabilities. The transformation of putting deeds behind the words has, however, been far from impressive. Notably, the explicitly announced ambitions regarding contributions to the international peace support operations decreased significantly once Bulgaria's memberships in NATO/EU were secured. Finally, regarding the ways, we find the government's argumentation for a *multilateral approach* trustworthy. Even if the government mainly has related this approach to the international context, we argue that the lack of means for projecting especially air power in the national context has made multilateralism a necessity rather than an option. However, we do not exclude a more bilateral approach focusing on the US also in this regard. When it comes to means and ways, Bulgaria's modest ambitions corresponds well to our expectations of a small state that is not strategically exposed.

Table 4.2 Bulgarian strategy.

Alignment strategy	Military strategy		
	Ends	Means	Ways
Multiple-courting	Survival	Expeditionary warfare	Multilateral approach

We agree with Blagovest Tashev when concluding that Bulgaria's transition regarding its defence policy was not initially guided by the goal of achieving membership in NATO and the EU. We also agree with his argument that the lack of national consensus may have delayed the steps towards the memberships. We note that already in 2004 he predicted that the political elite of Bulgaria, once the country became a member, could conclude that the memberships themselves provided security, which in turn could lead to conserving the traditional and regionally oriented profile of the armed forces. Clearly, different domestic ideological preferences remained important also after Bulgaria's entrance into the Alliance (Tashev 2004; see also Cooper and Oliver 2017). Like Tashev, Nikolay Dotzev stresses the importance of historical precedents. He argues that Bulgaria had begun to act as an ally to NATO long before it was invited to become a member (Dotzev 2005; see also Ratchev 2005). Margarit Tenev Mihaylov also places emphasis on Bulgaria's shortage of resources and finances. Even if we agree with her conclusions regarding Bulgaria's need to balance its security policy, we consider this relevant only immediately after the disintegration of the Warsaw Pact. Since this left Bulgaria alone to face Greece and Turkey, that is the two states Bulgaria was planning to confront in the event of a military conflict during the Cold War, the Bulgarian government had to find new solutions to the country's security challenges.

Clearly, the lack of external security guarantees and the regional military imbalances became great concerns. Arguably, hedging rather than balance of power has been the preferred alignment strategy at least from 1997 and onward. However, the shortage of resources and finances, in combination with the need of external security guarantees, may very well lead to a bandwagoning strategy with the US (Tenev Mihaylov 2013; see also Tashev 2004). Georgi Tzvetkov conclude that not only the Bulgarian economy but also Bulgaria's historical distrust for Turkey, as well as strong pro-Russian affiliations, provide explanations to the strategic standstill from the end of the Cold War until 1997. We find interesting his additional explanation that the democratically elected politicians strived to establish civilian control over the armed forces during these years. We agree with his findings that balance of power and even neutrality between NATO and Russia very well may have been the implicit policy until 1997. We agree with his conclusion that the transformation of the armed forces focused on expeditionary warfare but that a true modernisation, due to insufficient funding, never took place. Arguably, the traditional pro-Russian position can explain why a bandwagoning strategy with the US has not been put on the table, yet (Tzvetkov 2014; see also Hadjitodorov and Sokolov 2018; Wezeman and Kuimova 2018). We find Maria Neykova's additional explanation for the lack of modernisation, that is the governmental body's neglect of taking part of and contributing to the implementation of doctrinal documents in the area of national security, is plausible (Neykova 2017).

Notes

1 We would like to express our gratitude to First Secretary Anta Antoaneta Grigorova, Embassy of Bulgaria, Stockholm, and Lieutenant Colonel Rickhard Nordfjäll, Embassy of Sweden, Sofia for their support.
2 Ivan Kostov of the Union of Democratic Forces (UDF) served as prime minister 21 May 1997–24 July 2001. Simeon Borisov von Saxe-Coburg-Gotha, former Tsar Simeon II, of the National Movement (NM) served as prime minister 24 July 2001–17 August 2005. Sergei Stanishev of the Bulgarian Socialist Party (BSP) served as prime minister 17 August 2005–27 July 2009. Marin Raykov served as independent prime minister 13 March–29 May 2013. Plamen Oresharski served as independent prime minister 29 May 2013–6 August 2014. Georgi Bliznashki served as independent prime minister 6 August–7 November 2014. Ognyan Gerdzhikov of the National Movement for Stability and Progress (NMSP) served as prime minister 27 January–4 May 2017. Boyko Borisov of the Citizens for European Development of Bulgaria (GERB) served as Bulgaria's prime minister 27 July 2009–13 March 2013, 7 November 2014–27 January 2017 and 4 May 2017–12 May 2021. As of 1 January 2022, Brigadier General Stefan Yanev had served as independent prime minister since 12 May 2021. See www.gov.bg/en.

Bibliography

Bulgarian Council of Ministers (CM) (2002). *White Paper on Defence*.
——— (2008). *Updated Plan for Organizational Build-Up and Modernization of the Armed Forces*.
——— (2011). *National Defense Strategy*.
——— (2012). *Annual Report on the Status of Defence and the Armed Forces of the Republic of Bulgaria 2011*.
——— (2014). *Vision: Bulgaria in NATO and in European Defence 2020*.
——— (2015). *Programme for the Development of the Defence Capabilities of the Bulgarian Armed Forces 2020*.
——— (2017). *National Plan for Increasing the Defense Spending to 2% of the Gross Domestic Product Until 2024*.
Bulgarian Ministry of Defence (MoD) (2010). *The Republic of Bulgaria's Armed Forces' Development Plan*.
Bulgarian National Assembly (NA) (2010a). *Law on Defence and Armed Forces of the Republic of Bulgaria*. Decided on 26 February.
——— (2010b). *White Paper on Defence and the Armed Forces of the Republic of Bulgaria*. Adopted on 28 October.
——— (2011). *National Security Strategy of the Republic of Bulgaria*. Approved on 25 February.
Cooper, Harry and Christian Oliver (2017). 'Bulgaria Caught between NATO and the Kremlin' *Politico*, 17 February.
Crampton, Richard (2007). *Bulgaria*. Oxford: Oxford University Press.
Curtis, Glenn (1993). *Bulgaria: A Country Study*. Washington, DC: Library of Congress.
Dotzev, Nikolay (2005). 'The Soviet Legacy: Transforming Bulgaria's Armed Forces for Homeland Security Missions' *Connections* Volume 4, Issue 3.
Hadjitodorov, Stefan and Martin Sokolov (2018). 'Blending New-Generation Warfare and Soft Power: Hybrid Dimensions of Russia-Bulgaria Relations' *Connections* Volume 17, Issue 1.
International Institute for Strategic Studies (IISS) (2000). *The Military Balance 2000–2001*. Oxford: Oxford University Press.
——— (2010). *The Military Balance 2010*. London: Routledge.
——— (2020). *The Military Balance 2020*. London: Routledge.
Liddell-Hart, Basil (1997). *History of the Second World War*. London: Papermac.
Neykova, Maria (2017). 'Is Bulgaria's Legislation Up to the Mark to the Contemporary Security Environment?' *Globalization, the State and the Individual* Volume 14, Issue 2.
Ratchev, Valeri (2005). 'Defence Diplomacy: The Bulgarian Experience' in Timothy Edmunds and Marjan Malesic (eds). *Defence Transformation in Europe: Evolving Military Roles*. Amsterdam: IOS Press.
Tashev, Blagovest (2004). 'In Search of Security: Bulgaria's Security Policy in Transition' *Papeles del este*, Number 8.
Tenev Mihaylov, Margarit (2013). *Bulgaria in the Current Geopolitical Situation*. Carlisle, PA: US Army War College.
Tzvetkov, Georgi (2014). 'Defense Policy and Reforms in Bulgaria since the End of the Cold War: A Critical Analysis' *Connections* Volume 13, Issue 2.
Wezeman, Siemon and Alexandra Kuimova (2018). 'Bulgaria and Black Sea Security', *SIPRI Background Paper*, December.

5 The strategy of Croatia[1]

The primary sources analysed in this chapter consist of the strategic defence reviews of 2005 and 2013, the plans for the development of the Croatian armed forces of 2006 and 2014, the NSSs of 2002 and 2017, the 2006 act on the security intelligence system and the national cyber security strategy (NCSS) of 2015.[2]

5.1 Historical background

The Kingdom of Croatia was established in 925. Competition with Venice for control over the eastern Adriatic coast, as well as conflicts with Bulgarians, Byzantines and Hungarians, characterized the Kingdom's existence. In 1091, the death of King Stephen II led to the War of the Croatian Succession and a Hungarian invasion. In 1102, the Croatian crown was passed to the king of Hungary. Because of the Ottoman invasions and the collapse of the Hungarian kingdom, the Croatians elected Ferdinand I of the Habsburg dynasty as their new king in 1527. Following the defeat of Austria-Hungary in WWI, Croatia came to enter into a union forming the Kingdom of Serbs, Croats and Slovenes, renamed in 1929 the Kingdom of Yugoslavia. The renaming was an attempt to calm the domestic disturbances among the different ethnic groups, mainly the Croats and Serbs. However, in October 1934, an assassin with connections to the Croatian extreme nationalist *Ustaše* organisation killed the king. The political power was assumed by a governorship, the Yugoslav regency, which was in effect until March 1941. The regency could not prevent the internal tensions from escalating. Following the invasion of Yugoslavia by Nazi-Germany in April 1941, the Germans created a Nazi-controlled puppet state of Croatia. However, parts of the country were annexed by German allies Hungary and Italy. Notably, the UK not only hosted the Yugoslavian exiled regime but also supported the resistance. After liberation by communist partisans, the monarchy was abolished, and the country became a federal republic with a communist government. In June 1991, Croatia declared independence from Yugoslavia. The Croatian War of Independence lasted until the Erdut Agreement was reached on 12 November 1995 (see, for example, Liddell-Hart 1997; Goldstein 1999; Bartlett 2003).

DOI: 10.4324/9781003298052-7

5.2 Strategic environment

'The world increasingly faces new forms of threats', Croatian Minister of Defence Berislav Rončević concluded in 2005. 'International terrorism, smuggling of narcotics, weapons and human beings and the proliferation of weapons of mass destruction create enormous challenges for most states, thus becoming global concerns', he continued (Croatian MoD 2005:5). The assessments of the government led to the conclusion that a direct military threat to Croatia was highly unlikely the

> likelihood of a conventional conflict – in which Croatia's territory would be part of a larger battlefield or area of hostilities in the coming years – is very low. A likely aggressor that could potentially threaten Croatia's security would do so with conventional means and is not expected to possess highly sophisticated military capabilities.
> (Croatian MoD 2005:13)

'The probability of a conventional attack on the Republic of Croatia is very low', the government concluded also in 2006 (Croatian MoD 2006:8). The government did not, however, rule out the risks of a violent overthrow of the state authority structures (Croatian Parliament 2006). Threats such as terrorism, proliferation of WMD, crisis hotspots around the world, organized crime, increased abuse of cyberspace, climatic and demographic changes, as well as the increased risks of ecological, technical or technological accidents were elaborated by the Croatian government in 2013. The focus in the considerations was on the increased global energy needs and on the global economic crisis. The government observed that the latter had led to a decrease in the defence budget of not just Croatia. On the regional level, the government warned about the risks of political radicalizations without mentioning any specific actor or state. 'The Republic of Croatia is not facing, now nor in the foreseeable future, a direct threat of armed aggression on its territory', the government once again concluded (Croatian MoD 2013:13).

On the one hand, the government had, due to the economic crisis, to reduce the funds allocated for defence. On the other hand, the government noticed that the security environment at the global as well as at the regional level generated new security challenges and threats. These risks were exemplified by international terrorism, trans-border organized crime, proliferation of WMD, disruptions in the energy supply systems, the potential dangers from missile and cyber attacks and piracy. Instabilities arising in the close vicinity of the allies' borders could, the government concluded, 'have far-reaching impacts on the global security architecture, thus reaffirming the importance of a balanced approach to the Alliance's core tasks'. Numerous conflicts and crisis areas, emerging failed and unstable states and various types of threats caused by the rise of extremist ideologies imposed, the government stressed, 'the need for the development of key capabilities'. On the one hand, the government observed the risks that Croatia indirectly could be exposed to implications of potential regional conflicts. On the other hand, it concluded that there was no danger from direct, armed aggression on Croatian territory (Croatian MoD 2014:9). In addition, cyberterrorism and other cyber aspects of national security were given specific attention (Croatian Government 2015). 'The European area, in particular the area of Central Europe, to which Croatia belongs geographically, historically and culturally, is faced with security threats from the east and the south', Croatian President Mrs. Kolinda Grabar-Kitarović argued in 2017. Andrej Plenković, the Croatian prime minister, warned against 'increased and more complex security threats and risks' (Croatian Government 2017:4–5). Geopolitical competition among the major powers, the weakening of multilateralism, the outbreak of inter- as well as intra-state conflicts, Islamic radicalism, climate changes, and large migratory movements were all used to exemplify this complexity.

5.3 Ends

In 2002, the Croatian government claimed that its security strategy 'is characterized by its determined effort to advance towards and enter Euro-Atlantic and European security organisations' (Croatian Parliament 2002:1). Three years later, the government presented a list of national interests and objectives. The interests were labelled vital and important. The former category included the sovereignty, independence and territorial integrity of Croatia, intact national identity and values, and the protection and safety of Croatia's citizens and their property. The latter category included the preservation and development of democratic institutions, the rule of law, economic prosperity and social justice. Other important interests were a peaceful and secure international environment, an international order based on the principles of justice, as well as the preservation and protection of the natural environment. The objectives were labelled as either broad national objectives or specific security objectives. The former focused on creating proper conditions for the political, economic and social development of Croatian society. The latter group focused on the building of favourable regional and global security surroundings through international security organisations as well as assuring the national resources necessary to respond to all Croatia's security challenges, risks and threats. 'A consistent national and strategic objective has been to join NATO and the EU', the government announced (Croatian MoD 2005:12).

Protecting the sovereignty of Croatia, defending Croatia and its allies, participating in crisis-management operations abroad, participating in confidence- and security-building activities and assisting Croatian civilian institutions were the key objectives presented to the Croatian armed forces in 2006 (Croatian MoD 2006). Within the framework of the objective to defend Croatia and its allies, the government made clear the Croatian armed forces had to be able to deter, halt and repel an armed aggression against the country (Croatian MoD 2013). This was to be achieved independently, as well as with the help of the allies. The armed forces were also to be ready to contribute to the defence of the allies (Croatian MoD 2014). Like any other country, 'the security of its citizens and its national territory' is fundamental also for Croatia, Prime Minister Andrej Plenković argued in 2017. 'Values created in the victorious Homeland War are the fundaments on which a

64 *The empirical exploration*

modern Croatian security system is developed', he continued as he stressed the importance of a national Croatian identity and of the protection of Croats living outside the borders of Croatia. He especially emphasised the importance of protecting the latter's 'equality and constitutionality in neighbouring Bosnia and Herzegovina' (Croatian Government 2017:5). The safety of Croatia's citizens and the territorial integrity and sovereignty of Croatia were expressed as the basic preconditions for the existence of the Croatian state. The well-being and prosperity of the Croatian citizens, the stability of the Croatian democratic political system and society, as well as Croatia's international reputation and influence were also considered as fundamental ends. 'Equality, sovereignty and existence of the Croatian people in Bosnia and Herzegovina, as well as the position of Croatian national minorities in other countries and the Croatian diaspora enjoy special care and protection provided by Croatia', the government declared (Croatian Government 2017:8). The area of Southeast Europe was considered being of specific strategic importance and a priority in terms of security.

5.4 Means

'The armed forces will become fully professionalized and consist of a newly styled "contract" reserve. Military capabilities that enable Croatia to participate in international operations will be particularly stressed', the Croatian government announced in 2005 (Croatian MoD 2005:6). Overall force downsizing, including cutting the number of units and command headquarters, as well as suspending the obligatory conscription aimed at establishing a fully professionalized armed forces of 16,000 troops and additional 8,000 contracted reserves. Croatian armed forces must have the capacity to 'take part in a collective defence operation with pre-designated forces' as 'part of NATO Response Force (NRF) or EU Battlegroups', the government stressed. Until 2009, this capacity was defined as a reinforced company and after 2009 as a battalion (Croatian MoD 2005:18). Regarding the development of the army, the government concluded:

> The Land Forces will consist of two main units, a motorized brigade and an armoured-mechanized brigade [. . .]. Heavy armour is unlikely to be deployed [abroad]. The emphasis for deployment will be on light mobile forces that are sustainable and able to protect themselves. As a result, new acquisitions, such as highly mobile combat vehicles (on wheels), will be required.
> (Croatian MoD 2005:20)

The bulk of the navy was to consist of three fast missile crafts and land-based missile batteries. Inshore and port mine hunters were given priority. Regarding the air force, the government concluded that 'Croatia does not need to develop large-scale air defence capabilities'. Twelve MiG-21 aircraft were hence considered a relevant strength. However, the government also stressed the role of the air force supporting the operational needs of the land forces regarding deployment and sustainability as well as providing necessary infrastructure to NATO partners in case of need. The acquisition of two additional transport aircraft with medium- to long-range capability was given priority (Croatian MoD 2005:23). However, new combat aircraft for the air force and a new naval patrol ship for the navy were given priority in the long-term plan presented one year later. By 2010, the Croatian armed forces were to be manned exclusively on a volunteer basis, the government declared. The territorial principles that so far had guided the structuring of the armed forces were to be abandoned. At the same time, the government stressed the difficult economic situation and the limited financial resources. Consequently, the Croatian armed forces 'will not be able to develop the full spectrum of military capabilities necessary for independent action outside Croatian territory', the government made clear (Croatian MoD 2006:7–8). High readiness was the focus of the developments. Rapid deployment forces were hence to be ready for interventions within 24 hours. A reinforced mechanised company, a special forces company, a naval infantry unit and two helicopters were, over time, to constitute the bulk of the Croatian high-readiness forces. In addition, Croatian forces included in NRF or EU Battle Groups (EUBGs) 'will stand ready to deploy within 5 days', the government promised (Croatian MoD 2006:8).

Moreover, within 90 days, a special forces battalion, a mechanised battalion, a naval infantry company, two engineer companies and some combat service support units were to be ready to deploy internationally. The government declared its ambition simultaneously being able to contribute with troops to two international operations with forces equivalent to a reinforced battalion. The key equipment of the army included 267 MBTs out of which 192 were considered surplus. The navy sailed, amongst other vessels, three missile ships and four patrol ships. All 12 MiG-21 combat aircraft and seven Mi-24 helicopters of the air force were to be replaced. The lack of submarine warfare capability was, however, consciously neglected (Croatian MoD 2006). 'The armed forces were, the government announced in 2013, to undergo a gradual downsizing to 15 000 personnel by the end of 2017. The army was to consist of two guard brigades, a military police regiment, a signals regiment, a military intelligence battalion, and a battalion for chemical, biological, radiological and nuclear (CRBN) defence as well as artillery, engineer and logistics units. The navy was to consist of a flotilla of combat ships. Capabilities for anti-surface and anti-mine warfare as well as limited anti-submarine activities were to be developed. The air force was to keep the MiG-21 in operative use at least until 2019. The combat aircraft were primarily tasked to control and protect the national air space. The An-32 transport aircraft were to be withdrawn from operational use and acquisition of new transport aircraft was to be considered only after 2020. Without rotation, eight helicopters were to be deployable for a period of up to six months but only after 2018. On a rotation basis, half the number were to be deployable for longer periods. Clearly, the army had to bear the burden of international troop contributions. Croatia 'will be able to deploy forces the size of a battle group, based on one mechanized battalion, on a rotation basis, to a peace support operation', the government announced (Croatian MoD 2013:17).

66 *The empirical exploration*

Table 5.1 Main military resources of Croatia.

	2000	2010	2020
Army brigade	12	2	2
MBT	305	261	75
ACV/APC/IFV	127	141	300
Combat aircraft	41	12	11
Principle surface combatant	8 missile and patrol craft	2 corvettes, 7 patrol craft	5 patrol craft

Source: International Institute for Strategic Studies (2000, 2010, 2020).

'In order to respond to modern challenges and threats, the Armed Forces should be capable of performing non-traditional tasks. They should be adjustable, mobile, interoperable, deployable and sustainable in the Area of Operations', the government concluded in 2014 (Croatian MoD 2014:9). The government signalled continued contributions to the NRF as well as to the EUBG but admitted that the defence budget could not afford a participation with more than 200 troops. The key material was to include 48 MBT, five coastal patrol boats and two mine hunters. Regarding the air force, the government announced that the life cycle of the MiG-21 combat aircraft was to expire by the end of 2024 and that the priority hence was 'to find a permanent solution for protection of the air space' (Croatian Government 2017:73). See Table 5.1.

5.5 Ways

'Croatia's foreign policy will guide a robust program of international defence co-operation', the government declared in 2005 (Croatian MoD 2005:7). The role of the Croatian armed forces to support the country's foreign and security policy was emphasised. A continuation of Croatian contributions to UN-led military operation was high on the agenda, although achieving full Euro-Atlantic integration by joining NATO and the EU was prioritised. Several bilateral arrangements within the Euro-Atlantic frames were elaborated such as the cooperation between Croatia Hungary, Italy and Slovenia in developing a brigade-sized Multinational Land Force (MLF) unit designed for international peace support operations. The government presented a list of operations the Croatian armed forces were tasked to be able to conduct. Host nation support (HNS), deterring potential aggressors, a full spectrum of international crisis response operations and operations in order to defend Croatia and, although the country was not yet a member, NATO allies were on the list. Croatian armed forces 'are more likely expected to take part in operations abroad rather than at home [. . .]. Croatia will most likely participate in operations as part of a larger UN, NATO or EU force, or some other ad hoc coalition of forces', the government concluded (Croatian MoD 2005:17–18). The Croatian armed forces were to play a much more significant role abroad than previously, the government announced in 2006. 'Such actions will always be carried out within the frameworks of allied forces, whether NATO, the UN, EU or other coalition forces', the government declared (Croatian MoD 2006:7). Seven years later, Croatia had become a member in NATO as well as in the EU, but the message remained similar:

> Through participation in the international peacekeeping operations and missions under the leadership of NATO, UN, and the EU, by contributing forces or other forms of co-operation and development of the Armed Forces and regional military cooperation, the Republic of Croatia established itself as a responsible and credible member of the international community, as an ally fulfilling its obligations and contributing to security at the regional and the global level.
>
> (Croatian MoD 2013:2)

Croatia will continue to be a part of 'the wider, all-encompassing response by the international community, primarily within the UN, NATO and the EU', the government pledged (Croatian MoD 2013:13). The focus of Croatia's international defence cooperation had both a multilateral and a bilateral approach. In the former case, the cooperation was mainly to take place within the framework of NATO and the EU and, in the latter case, with member countries of those organizations as well as with countries and organizations in Croatia's region. 'An effective response to a broad spectrum of global threats is possible only through a strong and comprehensive cooperation with the allies and partners' the government concluded also in 2014 (Croatian MoD 2014:9). When contributing to international security efforts, the Croatian armed forces were to be capable of participating in a range of peace support operations, including defence diplomacy, arms control, disarmament and prevention of proliferation of WMD. The focus on contributions to the NRF and the EUBG 'does not exclude the possibility of joining other multinational formations' with the intention of immediate responses to and the management of international crisis situations, the government informed (Croatian MoD 2014:17). While the Croatian army as a whole was to be able to conduct joint defensive and offensive operations in order to defend Croatia's territorial integrity, its two brigades were given different tasks. Hence, the mechanized guard brigade was tasked to participate in international operations with a battalion-sized battle group on a rotation basis. The armoured guard brigade was to primarily develop capabilities national operations.

5.6 Conclusions: Croatian strategy

We find it reasonable to conclude that Croatia has adopted a *multiple-courting*, that is a hedging, alignment strategy. Clearly, memberships in both NATO and in the EU were prioritized in the early years of the new millennium. In addition, Croatia explicitly declared its willingness to cooperate not only with these the two organisations but within the frames of the UN and OSCE, as well as on ad hoc basis with different coalitions. Arguably, the experience of the Balkan wars may have prevented a bandwagoning strategy with the US becoming a trustworthy

68 *The empirical exploration*

alternative. When it comes to the ends of the military strategy, we conclude that *survival* is the most relevant label. This does not only go for the Croatian state but for the Croatian nation as well. Consequently, the vital interests consisted both of aspects of the state, that is sovereignty, independence and territorial integrity, and of the nation, that is the national identity and values. Well-being and prosperity were not only limited to Croatian citizens but also included the Croatian diaspora in Bosnia and Herzegovina and elsewhere. Regarding the means, the Croatian government has announced its priority on *expeditionary* capabilities. The transformation towards such a force structure has, however, been far from rapid. Despite acknowledging the immediate need of new combat aircraft for the air force and a new naval patrol ship for the navy, these two services have simply been unfit for contributing to expeditionary ambitions. Consequently, the army has had to bear the burden in this regard. Since the army for several years kept almost 200 MBTs as surplus, the necessary funding for a more compelling development of an expeditionary force has been delayed. The ambitions regarding participation in international peace support operations seems to have decreased since Croatia was admitted to NATO/EU. We find, regarding the ways, that the government's argumentation for a *multilateral approach* convincing. Although the government mainly has referred to the international context, we argue that the lack of means for projecting air and maritime power in the national context has made this approach a question not solely of preference but of necessity.

Table 5.2 Croatian strategy.

Alignment strategy	Military strategy		
	Ends	Means	Ways
Multiple-courting	Survival	Expeditionary warfare	Multilateral approach

We agree with Jelena Grčić Polić that the decrease in Croatia's defence budget over the years probably can be explained by other priorities and needs of post-war Croatian society and that joining both NATO and the EU hence became top priorities soon after the end of the war in 1995 (Grčić Polić 2003). Ryan Hendrickson and Ryan Smith present interesting conclusions when arguing that the presidency of Franjo Tudjman, from 1995 to 1999, was characterised by the politicisation of the armed forces in order to ensure that his party, the HDZ, remained in power. According to Hendrickson and Smith, this policy prevented Croatia's integration with the West, We agree when they conclude that after Tudjman's death in 1999, 'Croatia has made significant strides in reforming its military to comply with NATO standards' (Hendrickson and Smith 2006:302; see also Kornfein 2010; Rubisa 2009; Seroka 2008). We agree with Nathan Polak and his colleagues when arguing that, after gaining its NATO membership, Croatia has strived for creating a 'niche' role for itself in territorial defence. We find their arguments that Croatia's military purchases indicates 'a heavy emphasis on land-based equipment, perhaps indicative of a residual fear of regional conflict' compelling (Polak *et al.* 2009:508).

We do not, however, agree that the purchase of two ex-Finnish Helsinki-class missile ships necessarily reflects a Croatian desire to build offensive naval capabilities. We rather agree with their observation regarding the overall small size of the Croatian navy. Arguably, Amadeo Watkins provides support for our arguments on the failure of policy delivery regarding the fighter aircraft acquisition (Watkins 2019). We find Anita Perešin's observations regarding the formulation of Croatia's first Membership Action Plan (MAP) taking place in the aftermath of 9/11 and the focus on expeditionary warfare convincing (Perešin 2013). As Perešin, Drazen Smiljanic observe the context in which the development of strategic documents take place. We find noteworthy his observation that the development of Croatia's national security strategy presented in 2017 took 'place in a radically different security environment' compared to the previous strategy published in 2002. Clearly, his arguments on national defence and the focus on hybrid warfare are interesting and reflects the potential inconsistency regarding the official focus on expeditionary warfare (Smiljanic 2017). We agree with Zvonimir Mahečić when arguing that Croatia's strategic documents, on the one hand, tend to be quite optimistic in enumerating all sorts of military missions and military tasks while, on the other hand, it 'is quite obvious that Croatia does not have the necessary resources to provide for all these missions' (Mahečić 2010:72–73). Consequently, the Croatian military strategy risks being imbalanced, that is the means simply do not support the ends and ways. Arguably, the findings of Rebecca Cruise and Suzette Grillot regarding Croatia's ambition not only joining NATO and the EU but also striving for a seat in the UN Security Council (UNSC) support our conclusions on the alignment strategy based on hedging (Cruise and Grillot 2010). Arguably, the findings of Robert Barić and Dražen Smiljanić regarding US–Croatian relations also support our position, that is not labelling the Croatian strategy as bandwagoning (Barić and Smiljanić 2019).

Notes

1 We would like to express our gratitude to Lieutenant Colonel Rickhard Nordfjäll, Embassy of Sweden, Zagreb for his support.
2 Zlatko Mateša of the Democratic Union (HDZ) served as prime minister 7 November 1995–27 January 2000. Ivica Račan of the Social Democratic Party (SDP) served as prime minister 27 January 2000–23 December 2003. Ivo Sanader of HDZ served as prime minister 23 December 2003–6 July 2009. Jadranka Kosor of the HDZ served as prime minister 6 July 2009–23 December 2011. Zoran Milanović of the SDP served as prime minister 23 December 2011–22 January 2016. Tihomir Orešković served as independent prime minister 22 January–19 October 2016. As of 1 January 2022, Andrej Plenković of the HDZ had served as prime minister since 19 October 2016. See https://vlada.gov.hr/en.

Bibliography

Barić, Robert and Dražen Smiljanić (2019). 'Relations between the United States and Croatia: Development and Future Perspectives' in Anna Péczeli (ed). *The Relations of Central European Countries with the United States*. Budapest: Dialóg Campus.

Bartlett, William (2003). *Croatia: Between Europe and the Balkans*. London: Routledge.
Croatian Government (2015). *National Cyber Security Strategy*.
——— (2017). *National Security Strategy*.
Croatian Ministry of Defence (MoD) (2005). *Strategic Defence Review*.
——— (2006). *Croatian Armed Forces Long Term Development Plan 2006–2015*.
——— (2013). *Strategic Defence Review*.
——— (2014). *Croatian Armed Forces Long Term Development Plan 2015–2024*.
Croatian Parliament (2002). *Strategy for the Republic of Croatia's National Security*.
——— (2006). *Act on the Security Intelligence System of the Republic of Croatia*.
Cruise, Rebecca and Suzette Grillot (2010). 'The Development of Security Community in Croatia: Leading the Pack' *Croatian International Relations Review* Volume 16, Issue July–December.
Goldstein, Ivo (1999). *Croatia: A History*. London: Hurst & Company.
Grčić Polić, Jelena (2003). 'Security and Defence Reforms: A Croatian Armed Forces Case' *Croatian International Relations Review* Volume 9, Issue January–June.
Hendrickson, Ryan and Ryan Smith (2006). 'Croatia and NATO: Moving toward Alliance Membership' *Comparative Strategy* Volume 25, Issue 4.
International Institute for Strategic Studies (IISS) (2000). *The Military Balance 2000–2001*. Oxford: Oxford University Press.
——— (2010). *The Military Balance 2010*. London: Routledge.
——— (2020). *The Military Balance 2020*. London: Routledge.
Kornfein, Iva (2010). 'Concept of Security in Croatia and the European Security Strategy' *Croatian International Relations Review* Volume 16, Issue January–June.
Liddell-Hart, Basil (1997). *History of the Second World War*. London: Papermac.
Mahečić, Zvonimir (2010). 'Security Policies in the Western Balkans: Croatia' in Miroslav Hadžić, Milorad Timotić and Predrag Petrović (eds). *Security Policies in the Western Balkans*. Belgrade: Belgrade Centre for Security Policy.
Perešin, Anita (2013). 'Croatian Counter-Terrorism Strategy: Challenges, Prevention and Response System' *Research Paper No. 160*. Athens: Research Institute for European and American Studies.
Polak, Nathan, Ryan Hendrickson and Nathan Garrett (2009). 'NATO Membership for Albania and Croatia: Military Modernization, Geo-Strategic Opportunities and Force Projection' *Journal of Slavic Military Studies* Volume 22, Issue 4.
Rubisa, Damir (2009). 'The Europeanization of Croatia's Security Discourse' *Politicka Misao: Croatian Political Science Review* Volume 46, Issue 5.
Seroka, Jim (2008). 'Assessment of the Transformation of Civil-Military Relations in Serbia and Croatia since 2000' *Politicka Misao: Croatian Political Science Review* Volume 45, Issue 5.
Smiljanic, Drazen (2017). 'Development of the Croatian National Security Strategy in the Hybrid Threats Context' *Croatian International Relations Review* Volume 23, Issue 80.
Watkins, Amadeo (2019). 'Fighter Aircraft Acquisition in Croatia: Failure of Policy Delivery' *Defense & Security Analysis* Volume 35, Issue 3.

6 The strategy of the Czech Republic (Czechia)[1]

The primary sources analysed in this chapter consist of the strategic documents of the Ministry of Defence (MoD) issued 2002, 2004 and 2008, the DWP of 2011, the NDSs of 2012 and 2017, the NSSs of 2011 and 2015, and the plans for the long-term perspective on defence of 2015 and 2019. In addition, the doctrine of the Armed Forces of the Czech Republic (AFCR) of 2004 as well as the cyber security and defence strategies of 2015 and 2018 have been explored.[2]

6.1 Historical background

The Duchy of Bohemia was established in late ninth century as a principality. During the tenth century, the Duchy conquered Moravia and Silesia, that is the two other parts that form the modern-day Czech Republic. In the early eleventh century, the Duchy became an imperial state of the Holy Roman Empire. The Duchy was raised to the status of Kingdom in 1198, and the House of Habsburg inherited the crown in 1526. The kingdom remained a part of the Austrian Empire until the defeat of Austria-Hungary in WWI. The independence of Czechoslovakia was proclaimed on 28 October 1918. Following the Treaty of Saint-Germain-en-Laye in September 1919, Bohemia and Moravia, as well as parts of Silesia, were incorporated with the newly created state of Czechoslovakia. Among all the new states established in central Europe after WWI, only Czechoslovakia preserved a democratic government throughout the interwar period. In late 1938, following the Munich agreement, Nazi-Germany annexed Sudetenland, while Poland annexed the Zaolzie region. The remaining parts of the first Czechoslovakian republic was organised as the second republic. The Germans occupied the Czech part of this republic in March 1939; hence Czechia ceased to exist as an independent state. A government in exile was organised in the UK, and Czechoslovakian troops contributed to the war efforts of the Allies. During the final phase of WWII, the USSR occupied both parts of Czechoslovakia and later oversaw their reunification as the third Czechoslovak republic. Following the Czechoslovakian uprising against the communist regime and the invasion by WP forces in 1968, Czechoslovakia was reorganized as a federal republic consisting of a Czech and a Slovak republic. Three decades later, the so-called Velvet Revolution of 1989 led to a definite

DOI: 10.4324/9781003298052-8

end of the communist regime. In late November 1992, both federal republics agreed on the dissolution of Czechoslovakia as of 31 December 1992 (see, for example, Wallace 1977; Gorys 1996; Liddell-Hart 1997).

6.2 Strategic environment

'Occurrence of a large-scale conventional aggression against the Czech Republic and NATO in the foreseeable time frame is highly unlikely', the government concluded in 2002 (Czech MoD 2002:4). Security was nevertheless considered as threatened by terrorist attacks, the proliferation of WMD, religious and ethnic conflicts, organised crime, massive and illegal migration, drugs trafficking and the instability of the political order. 'These risks may lead to limited armed conflicts, with a possibility of spreading', the government warned (Czech MoD 2002:5). 'The United States remain the most important member of NATO, which is the chief guarantor of the Czech Republic's security', the government declared in 2004 after joining the EU (Czech Ministry of Foreign Affairs (MoFA) 2004:20). The 'increasing tendency of asymmetric threats, augmented by possible WMD application, result [sic] in widening of the spectrum of the tasks' that the Czech security and defence policy will have to address, in addition to 'formulation of new priorities of the armed forces development', the government argued. 'Occurrence of large-scale armed conflict interfering [with Czech] territory and territory of another NATO member country in predictable time horizon is very improbable', the government nevertheless concluded (Czech MoD 2004:2).

Other serious threats could, the government argued, 'originate from regional crises caused by long-lasting unresolved problems that can escalate into intrastate or interstate armed conflicts'. Instability caused by failed or failing states, disruption in the flow of strategic resources including disputes over their control, abuse of information, humanitarian crises and natural disasters 'also require appropriate attention', the government concluded (Czech MoD 2008:3). The growing demand for key raw materials, the activities on financial markets and the unilateral attempts by some states to establish spheres of influence through a combination of political, economic and military pressure were considered threats to not only Czech but to European security as a whole, the governments argued in 2011. 'The probability of a direct threat to the territory of the Czech Republic by massive military attack is low', the government nevertheless concluded (Czech MoFA 2011:8). The government put emphasis on the political and economic stability of

> the environment in which the Czech Republic exists [as] the key external condition. More than three quarters of the Czech Gross Domestic Products (GDP) depend upon export and import. The extraordinary openness of the Czech economy renders the country dependant on the external environment, primarily in terms of market accessibility and availability of energy sources.
> (Czech MoD 2011:38)

Four years later, Prime Minister Bohuslav Sobotka observed that in 'the present world full of crises, the Czech Republic naturally has to face an enormous number of challenges'. Economic and social development was, the prime minister argued, 'our main and immediate concern. However, today there can be no development without a responsible approach to external security' (Czech MoFA 2015:3). The government stressed that the security of the Republic was 'crucially linked to the political and economic stability of the EU. As an exceptionally open economy, the Czech Republic is exposed to external influences particularly as regards access to markets and to energy sources' (Czech MoFA 2015:6). Given the overall decline of security and stability in Eastern Europe following the war in Ukraine, the government argued that it was 'impossible to entirely rule out a direct threat to the territories of some NATO and EU member states. Threats to the security of allies may be of the classical military nature or they may take the vague form of hybrid warfare' (Czech MoFA 2015:10). The government explicitly expected increased aggressive ambitions from regional superpowers including from Russia. 'The warning time for major conventional conflicts, even in the proximity of NATO and EU states, has decreased to weeks or months', the government warned and predicted 'an increased number of conflicts and a direct military threat to the Allies' (Czech MoD 2015:6). Regarding the risks of cyber attacks, the government feared that 'the Czech Republic may be targeted as a test bed for a major attack on our allies or states with a greater strategic importance and which use the same technologies, security mechanisms, and procedures' (Czech National Security Authority (NSA) 2015:11; see also Czech National Cyber Operations Centre (NCOC) 2018).

6.3 Ends

In 2002, the government presented two types of security interests. The vital interests focused on sovereignty, territorial integrity, principles of democracy, the rule of law, and the fundamental conditions for the life of the Czech Republic's citizens. The strategic interests were (1) increasing the defensive capabilities of the Czech Republic and its share in the defence of the allies, (2) participating in operations led by NATO, other international organisations and *ad hoc* coalitions, (3) reinforcing the regional cooperation as well as NATO's enlargement, (4) strengthening the security in the Euro-Atlantic region, (5) ensuring continued US presence in Europe, and (6) participating in the European Security and Defence Policy (ESDP). 'The main mission of the Armed Forces of the Czech Republic is defence of the Czech Republic and its security interest. Their basic task is preparation for defence of the Czech Republic and its defence against an external attack', the government declared (Czech MoD 2002:5). In addition to the other Visegrád countries, the bilateral relations with Austria and Germany were given priority. 'The development of good relations and close cooperation between the Czech Republic and the US and, within a broader framework, between the EU and the US, is one of the enduring priorities of the Czech foreign policy', the government declared. The influence and status not only of the Czech Republic but of the EU as a whole were central in the elaborations, not least in the economic context (Czech MoFA 2004:291).

The vital interests were slightly adjusted after the entrance into the EU. Instead of citizens, the government stressed the liberty of inhabitants. The strategic interests now also included increase of UN effectiveness, establishment of strategic partnership between NATO and the EU, complementary development of NATO and EU defence capabilities, development of the OSCE role in the prevention of armed conflicts, and ensuring the economic security of the Czech Republic (Czech MoD 2004). Notably, the armed forces focused on only two core objectives: defending the sovereignty and the territorial integrity of the Czech Republic and fulfilling its allied commitments (AFCR 2004). Establishing a reputation as a trustworthy and reliable ally was, in addition, mentioned as a strategic interest in 2011 (Czech MoD 2011). The 'defence of the state territory, its democratic system, citizens and their rights and freedoms' was the fundamental objective in the strategic elaborations one year later (Czech MoD 2012:7). Safeguarding the Czech Republic's economic security, strengthening the competitiveness of the Czech economy, as well as safeguarding the republic's energy, raw material and food security were all mentioned as strategic interests in 2015 (Czech MoFA 2015). Ensuring the security of citizens and protecting the sovereignty, territorial integrity, principles of democracy and the rule of law were described as the fundamental ends also in 2018 (Czech NCOC 2018).

6.4 Means

In 2002, the bulk of the Czech armed forces were to consist of one mechanised army division and one combined air force division. The government announced its decision to transform the armed forces from a conscription-based into a professional force of about 35,000 soldiers and 10,000 civilian employees. Regarding taking part in crisis response of higher intensity, the ambition was to contribute to one operation with a contingent equal to an army battalion of up to 1,000 troops. At the same time, a contingent of 250 persons was to be able to participate in a second, simultaneous operation of lower intensity for a short period of a maximum six months without rotation (Czech MoD 2002). The division task force, with a core created from a mechanized division, was also the focus of the elaborations in 2004. 'This task force will be assigned only on the CR territory or in its close neighborhood', the government declared (Czech MoD 2004:4). For operations under Article 5 of the Washington Treaty, Czech armed forces were to provide a brigade task force with up to 3,000 troops or a resource-equivalent task force from the air force. The 'Czech Republic will able to participate in one alliance, union or coalition operation of peace enforcement' with these alternative task forces, the government announced (Czech MoD 2004:4). In 2008, the government declared its readiness, in addition, to contribute with 'stand-by task forces up to the level of a battalion' for the NRF or EUBG but only 'when the brigade task force is not deployed'. When developing the armed forces, the focus shall, the government pledged, primarily be on 'strengthening of deployable capabilities' (Czech MoD 2008:5 and 7).

In 2011, the government presented a list of the key materiel of the armed forces including 166 MBTs, 494 infantry fighting vehicles (IFVs), 244 artillery pieces, 25 combat helicopters, and 38 combat aircraft. All numbers were far below the figures stated in the Treaty on Conventional Armed Forces in Europe and the Vienna Document. However, with the leasing deadline for JAS-39 Gripen combat aircraft approaching, the end of the service life of the surface-to-air missiles, the air defence radar equipment, the BMP-2 fighting vehicles and the Mi-24/35 attack helicopters, as well as the need for the modernisation of the artillery, the government concluded that the 'situation we are now in is critical' (Czech MoD 2011:13). The government concluded that the Czech armed forces should have up to 27,000 military personnel but admitted that the personnel strength temporarily had been decreased to 22,000 during the last couple of years despite the fact that the tasks had remained unchanged. The government made clear that the armed forces were to be ready to perform its tasks across the full spectrum of military operations and intensity scale. Specific attention was put on capacities needed for rapid and flexible response to unexpected events and on the ability to operate and sustain multiple concurrent operations. 'In respect of military capabilities development, an emphasis will be placed on the preparedness for warfare in urban areas and their immediate vicinity', the government declared (Czech MoD 2015:11).

In 2017, the government announced its ambitions to foster close defence cooperation with Germany, including the creation of multinational larger formations earmarked as follow-on forces for collective defence operations. The premise was that the Czech Republic would assign a brigade to a German army division. In addition, the Czech Republic was to enhance the trilateral cooperation with Germany and Poland by strengthening the capabilities of the Multinational Corps Northeast (MNC-NE). Based on the evaluation of the security threats, the government concluded that it was essential to develop new units in order to enhance the combat capabilities of especially the ground forces and thus increase the 'personnel numbers by additional 5,000 military professionals' (Czech MoD 2017:13). In order to ensure the defence of the republic, the government considered it vital to develop capabilities for operations in cyberspace. The cyber operations were to 'be carried out both within cyber defence of the Czech Republic and as part of military operations', the government declared (Czech NCOC 2018:7). In 2019, the government not only decided to exchange the mechanised brigade predesignated for contribution to collective defence operations for a heavy-type brigade to be available by 2026. The government also gave priority to the build-up of the already existing heavy brigade. For international crisis management operations, the Czech army was to be able to either deploy a brigade task force without rotation for a six-month period or simultaneously deploy both a battalion and a company task force with rotation. In addition, a rapid reaction battalion was ready to be deployed as a contribution to NRF or EUBG or within the framework of the Enhanced Forward Presence. Notably, regarding the collective defence operations, the government was referring to both Article 5 of the North Atlantic Treaty

76 *The empirical exploration*

Table 6.1 Main military resources of Czechia.

	2000	2010	2020
Army brigade	3	2	2
MBT	792	175	30
ACV/APC/IFV	1,204	622	248
Combat aircraft	110	48	38

Source: International Institute for Strategic Studies (2000, 2010, 2020).

and Article 42.7 of the Treaty on European Union. The core of the land force peacetime structure was to consist of two brigades, equipped with various types of combat vehicles including MBT, and one airborne infantry regiment. Regarding the air force, the government announced that the decision 'on either extended lease or acquisition of supersonic aircraft will be made no later than by 2025. At the same time, the potential increase of the strength of the supersonic fleet will be considered based on a security environment assessment and the related defence requirements review' (Czech MoD 2019:21).

6.5 Ways

In 2002, the government announced its ambition ensuring the defence of the Republic primarily by participating in international efforts abroad. An active engagement in crisis response operations outside the territory of the Republic, especially in operations led by NATO, the EU, the UN and the OSCE, or within *ad hoc* coalition groupings, was the first pillar of the Czech military strategy, the government declared. The protection and defence of Czech Republic airspace was the second pillar. The third pillar was the preparation and implementation of measures necessary for receiving allied reinforcements (Czech MoD 2002). The government declared its ambition developing both bilateral relations and participating in multilateral activities in order to promote Czech national interests. When addressing challenges related to the non-proliferation of WMD, terrorism and regional conflicts, the EU was said to provide the preferable framework. However, the government also 'emphasised the need to preserve the complementarity of EU and NATO' (Czech MoFA 2004:11). The government put emphasis on compatibility of the EUBG and NRF and declared its ambition to create a joint battle group with Austria and Germany. Regarding the situation in Afghanistan, the government announced that the Czech contributions included a mine clearing team to ISAF (International Security Assistance Force [in Afghanistan] and a contingent of 120 special forces personnel to Enduring Freedom, that is the US-led anti-terrorist operation. In addition, talks with Germany on a Czech contribution to the German-led PRT (Provincial Reconstruction Team) had been initiated. When it comes to the situation in the Balkans, 420 troops were serving in the joint Czech and Slovak KFOR (Kosovo Force) battalion (Czech MoFA 2004).

'In accordance with the current character of an international community involvement in peace-keeping operations, the Czech Republic Army will contribute to more running operations of less numerous contingents', the government announced. Effective preventive or deterrent actions decrease the necessity of reactive arrangements, the government argued (Czech MoD 2004:5). Four different contexts in which the Czech military was to be able to perform its duties were identified: (1) high-intensity collective defence operations; (2) low-intensity collective defence operations; (3) operations outside Article 5; and (4) military operations, as a tool of preventive diplomacy (AFCR 2004). 'The current security environment calls for an active involvement of the Armed Forces in operations outside the Czech Republic's, NATO's and EU's territories' the government concluded (Czech MoD 2008:4). The Republic prepares 'for active participation in NATO, EU and UN missions addressing the full range of crises – before, during and after conflicts', the governments informed in 2011 (Czech MoFA 2011:12). The AFCR 'can be deployed in all climatic zones, except for polar. Their deployment is not expected in naval landing operations', the government made clear (Czech MoD 2011:45). In order to maintain and develop defence capabilities, the government emphasised multinational solutions primarily through NATO's Smart Defence and the EU's Pooling and Sharing initiatives. The government prioritised 'the preservation of specific military and otherwise irreplaceable capacities of industrial production, and research and development in the Czech Republic', as well as to the Visegrád group regarding generating 'improvements in military capabilities as well as savings' (Czech MoD 2012:8).

Three years later, the government made clear that its 'security policy is based on a proactive approach' and that it strived 'to ensure that threats are detected early and analysed thoroughly, and that active measures are put into place'. The government also announced that it preferred 'active prevention of armed conflicts and the use of preventive diplomacy' (Czech MoFA 2015:6). The core elements of action were hence said to be active involvement in the NATO collective defence system, the development of EU crisis management capabilities, as well as cooperation with partner countries. 'The Czech Republic prefers multilateral approaches', the government announced (Czech MoFA 2015:16). 'The most likely and frequent example of the employment of the Czech Armed Forces in the next twenty years will be their participation in international operations outside the Czech Republic', the government predicted (Czech MoD 2015:8). The government announced its ambition intensifying military cooperation among the Visegrád countries in order to 'identify efficient solutions for the joint acquisition of military capabilities' (Czech MoD 2017:11).

6.6 Conclusions: Czech strategy

Arguably, the Czech Republic has adopted a *multiple-courting* alignment strategy that is quite balanced regarding the commitments to NATO and the EU, respectively. While the former organization is perceived as the ultimate provider of military security, the latter is considered relevant regarding not only military threats but other security challenges as well. The UN rather than the US is given a complementary role in the strategy. Even if the survival of the Czech state is elaborated, the government tends to put more emphasis on the economic dimension. Hence, we argue that gaining influence in and through the EU before Russia's illegal annexation of Crimea was a more prioritised end compared to survival against military threats. Since the war in Ukraine, *influence* and *survival* have been given similar attention and priority. Notably, the most important trade partner is also the ally the government is putting most emphasis on and developing closer military cooperation with: Germany. We find it reasonable to conclude that the Czech government has put deeds behind the words regarding transforming the military, that is the means, towards a force capable conducting expeditionary warfare.

However, since the war in Ukraine erupted in 2014, we argue that the means for national defence have been prioritized. It is noteworthy that the Czech armed forces strives to integrate with its German counterparts both when it comes to high-readiness and deployable forces for international missions and for more robust and heavy forces for collective defence operations. Arguably, the ambition to develop closer multilateral integration not only on the division level but also on the corps level that has been announced lately indicates increased attention on warfighting in the latter context. Obviously, the *multilateral approach* is the preferred way in both these contexts. This approach also guides the government regarding the force generation perspective. In addition to Germany, the other Visegrád countries are prioritised partners in this regard. Clearly, both the domestic defence industry and the national economic dimension are considered important by the government when developing cooperation on defence materiel, procurement and acquisition.

Table 6.2 Czech strategy.

Alignment strategy	Military strategy		
	Ends	Means	Ways
Multiple-courting	Influence/survival	National defence	Multilateral approach

We appreciate that Josef Procházka also is exploring the Czech defence strategy although with other definitions from ours. Nevertheless, we agree with him when arguing that the Czech strategic elaborations have been resource driven (Procházka 2015; see also Procházka and Dycka 2016). We also agree with Radek Khol when notifying the importance of the defence industrial aspects in formulating the Czech strategy. However, we do not agree that the Czech government has elaborated solely on two options in this regard, that is NATO-wide versus bilateral with the US. We rather argue that the government has favoured the EU/NATO, that is the European approach (Khol 2005; see also Dyčka and Procházka 2018; Procházka et al. 2018). Consequently, we find Juraj Marušiak providing interesting insights regarding the Visegrád aspects of the Czech hedging strategy (Marušiak 2015; see also Lazar 2019). Moreover, the Czech withdrawal from plans to participate in the US missile defence program clearly indicates that the US is not the key ally in the Czech strategy (Brusenbauch Meislová 2019; Dodge 2020; Pavlíčková and Bartoszewicz 2020).

Arguably, the findings presented by Josef Procházka in his previous research support our conclusions. Like Procházka, we have observed the successful integration of the Czech military into Western security structures and the contributions of the armed forces ensuring the Republic's recognition and influence in these structures. Moreover, like Procházka, we also observe that the military transformation has 'led to a flattened command structure and a deployable, sustainable, and fully professional force structure' (Procházka 2009:32). Zdeněk Kříž and Miroslav Mareš also provide support when noting the deployments of the Czech military to 'expeditionary operations under the auspices of various international organizations' (Kříž and Mareš 2011:60). Like Kříž and Mareš, Tomáš Weiss argues in line with our conclusions when claiming that other causes of the transformation than solely NATO membership has to be included in the elaborations (Weiss 2013; see also Procházka *et al*. 2016). We find compelling Miloš Balabán and his colleagues' arguments regarding the Czech government's preference to contribute to military missions with a strong peacekeeping and humanitarian dimension (Balabán *et al*. 2008; see also Hlouchová 2018). Consequently, we also find interesting the observations of Joseph Bell and Ryan Hendrickson regarding Czech reluctance to participate in NATO's mission. Notably, they connect this hesitation to Germany's opposition to the use of force, stressing the fact that the Czech government 'resisted pressure from the major powers in the alliance' including the US, the UK and France. Moreover, they concluded that the EU 'played virtually no meaningful role in this conflict' (Bell and Hendrickson 2012:160). Zdeněk Kříž and Martin Chovančík provide additional support for our conclusions when claiming that the main goal of Czech military reforms has been to establish military forces capable of expeditionary warfare not only within the framework of NATO but under the EU and the UN umbrellas as well (Kříž and Chovančík 2013).

Notes

1 We would like to express our gratitude to Lieutenant Colonel David Pastyrik and Major Zuzana Brázdová, Embassy of the Chech Republic, Tallinn, and Lieutenant Colonel Jörgen Marqardsen, Embassy of Sweden, Prague, for their support.
2 Miloš Zeman of the Social Democratic Party (ČSSD) served as prime minister 17 July 1998–12 July 2002. Vladimír Špidla of the ČSSD served as prime minister 12 July 2002–4 August 2004. Stanislav Gross of the ČSSD served as prime minister 4 August 2004–25 April 2005. Jiří Paroubek of the ČSSD served as prime minister 25 April 2005–4 September 2006. Mirek Topolánek of the Civic Democratic Party (ODS) served as prime minister 4 September 2006–8 May 2009. Jan Fischer served as independent prime minister 8 May 2009–28 June 2010. Petr Nečas of the ODS served as prime minister 28 June 2010–25 June 2013. Jiří Rusnok served as independent prime minister 25 June 2013–17 January 2014. Bohuslav Sobotka of the ČSSD served as prime minister 17 January 2014–6 December 2017. Andrej Babiš of the ANO 2011 served as prime minister 6 December 2017–28 November 2021. As of 1 January 2022, Petr Fiala of ODS had served as prime minister since 28 November 2021. See www.vlada.cz/en/.

Bibliography

Armed Forces of the Czech Republic (2004). *Doctrine of the Armed Forces of the Czech Republic*.

Balabán, Miloš, Jan Eichler, Josef Fučík, Vladimír Karaffa, Vladimír Krulík, Vladimír Kváča, Bohuslav Pernica and Antonín Rašek (2008). *Development of Security Policy and Strategy of the Czech Republic 1990–2009*. Prague: CESES Security Policy Centre.

Bell, Joseph and Ryan Hendrickson (2012). 'NATO's Visegrad Allies and the Bombing of Qaddafi: The Consequence of Alliance Free-Riders' *Journal of Slavic Military Studies* Volume 25, Issue 2.

Brusenbauch Meislová, Monika (2019). 'Relations between the United States and the Czech Republic: From Honeymoon to Hangover?' in Anna Péczeli (ed). *The Relations of Central European Countries with the United States*. Budapest: Dialóg Campus.

Czech Ministry of Defence (MoD) (2002). *Military Strategy of the Czech Republic*.
——— (2004). *Military Strategy of the Czech Republic*.
——— (2008). *Military Strategy of the Czech Republic*.
——— (2011). *White Paper on Defence*.
——— (2012). *Defence Strategy of the Czech Republic*.
——— (2015). *Long Term Perspective for Defence 2030*.
——— (2017). *Defence Strategy of the Czech Republic*.
——— (2019). *Long Term Perspective for Defence 2035*.

Czech Ministry of Foreign Affairs (MoFA) (2004). *Report on the Foreign Policy of the Czech Republic*.
——— (2011). *Security Strategy of the Czech Republic*.
——— (2015). *Security Strategy of the Czech Republic*.

Czech National Cyber Operations Centre (NCOC) (2018). *Cyber Defence Strategy 2018–2020*.

Czech National Security Authority (NSA) (2015). *Cyber Security Strategy 2015–2020*.

Dodge, Michaela (2020). 'Russia's Influence Operations in the Czech Republic during the Radar Debate' *Comparative Strategy* Volume 39, Issue 2.

Dyčka, Lukáš and Josef Procházka (2018). 'Czech Defence Policy Assessment in 2017/2018' *Vojenské rozhledy* Volume 29, Issue 4.

Gorys, Erhard (1996). *Czech and Slovak Republics*. London: Pallas Athene.

Hlouchová, Iveta (2018). *Czech Approach toward Counterinsurgency*. Brno: Masaryk University.

International Institute for Strategic Studies (IISS) (2000). *The Military Balance 2000–2001*. Oxford: Oxford University Press.
——— (2010). *The Military Balance 2010*. London: Routledge.
——— (2020). *The Military Balance 2020*. London: Routledge.

Khol, Radek (2005). 'Czech Republic: Prague's Pragmatism' *Contemporary Security Policy* Volume 26, Issue 3.

Kříž, Zdeněk and Martin Chovančík (2013). 'Czech and Slovak Defense Policies Since 1999: The Impact of Europeanization' *Problems of Post-Communism* Volume 60, Issue 3.

Kříž, Zdeněk and Miroslav Mareš (2011). 'Security Sector Transformation in the Czech Republic' *International Issues & Slovak Foreign Policy Affairs* Volume 20, Issue 3.

Lazar, Zsolt (2019). 'Success and Failures of the Gripen Offsets in the Visegrad Group Countries' *Defense & Security Analysis* Volume 35, Issue 3.

Liddell-Hart, Basil (1997). *History of the Second World War*. London: Papermac.

Marušiak, Juraj (2015). 'Russia and the Visegrad Group: More Than a Foreign Policy Issue' *International Issues & Slovak Foreign Policy Affairs* Volume 24, Issue 1–2.

Pavlíčková, Kristýna and Monika Gabriela Bartoszewicz (2020). 'To Free or Not to Free (Ride): A Comparative Analysis of the NATO Burden-Sharing in the Czech Republic and Lithuania' *Defense & Security Analysis* Volume 36, Issue 3.

Procházka, Josef (2009). 'The Defense Policy of Czechoslovakia and the Czech Republic since 1989: Stages, Milestones, Challenges, Priorities, and Lessons Learned' *Connections* Volume 8, Issue 2.

——— (2015). 'Adaptation of the Czech Republic Defence Policy: Lessons Learned' *Security & Defence Quarterly* Volume 6, Issue 1.

Procházka, Josef and Lukás Dycka (2016). 'Adaptation of the Czech Republic Defence Policy to the Dynamics of the Security Environment' *Strategic Impact* Volume 59, Issue 2.

Procházka, Josef, Lukáš Dyčka and Jakub Landovský (2016). 'Czech Defence Policy Response to Dynamics in Security Environment Development' *Vojenské rozhledy* Volume 25, Special Issue, September.

Procházka, Josef, Antonín Novotný, Richard Stojarand and Frank Libor (2018). 'The Long Term Perspective for Defence 2030: Comparative Analysis' *Politické vedy* Volume 21, Issue 4.

Wallace, William (1977). *Czechoslovakia*. London: Ernest Benn Ltd.

Weiss, Tomáš (2013). 'Fighting Wars or Controlling Crowds? The Case of the Czech Military Forces and the Possible Blurring of Police and Military Functions' *Armed Forces & Society* Volume 39, Issue 3.

7 The strategy of Estonia[1]

The primary sources analysed in this chapter consist of the national security concepts of 2004, 2010 and 2017, the NDS of 2011, the NCSSs of 2008, 2014 and 2019, as well as the defence development plans of 2009, 2013 and 2017. In addition, the national defence act of 2015, the defence industry policy of 2013 and the defence action plan of 2019 have been studied.[2]

7.1 Historical background

Following the crusade against pagans in Northern Europe, called for in 1193 by Pope Celestine III, southern Estonia and Latvia came to form Terra Mariana, controlled by the Livonian Order, while northern Estonia became the Danish Duchy of Estonia. In February 1918, after centuries of rule by Danes, German, Poles, Russians and Swedes, Estonia gained independence from Russia. However, the Estonians had to fight a war of independence against the Soviets before independence was recognized with the Treaty of Tartu in February 1920. In March 1934, the initially introduced democracy was, after a *coup d'état*, replaced by an authoritarian regime. In early August 1940, Estonia and the two other Baltic States were annexed by the USSR and became Soviet republics. Following the German attack on the USSR in June 1941, they were all occupied by Nazi-Germany as *Reichskommissariat Ostland*. In late 1944, the Red Army reoccupied and restored all three Baltic States as Soviet republics. Following the *coup d'état* attempt in Moscow in August 1991, Estonia successfully declared its restoration of full independence from the USSR (see, for example, Liddell-Hart 1997; Kasekamp 2010; Taylor 2018).

7.2 Strategic environment

According to the Estonian government, the enlargement of both NATO and the EU had contributed to reducing the threat of large-scale military conflicts in Europe. 'The globalisation of economic and social processes and communication is creating close links between nations and exerting a crucial effect upon the security environment', the government argued (Estonian Government 2004:4). Despite these positive developments, the international security environment was still not considered as having reached a state of stability. Threat of external coercion, instability or breakdown of information systems, local and regional crises, international terrorism, uncontrolled spread of WMD; organised crime; trafficking of weapons, narcotics and people as well as uncontrolled flow of refugees were all perceived as constituting threats to Estonia's security. The most likely sources of a military threat considered by the government included:

- the unexpected increasing or redeployment of military forces stationed near Estonia's border;
- large-scale military manoeuvres, which are not in compliance with international arms control treaties, in the direct vicinity of Estonia's border;
- the intentional violation of Estonia's air space, land border, or territorial waters.

(Estonian Government 2004:6)

'The asymmetrical threat posed by cyber-attacks and the inherent vulnerabilities of cyberspace constitute a serious security risk', the government argued following the coordinated cyber attack against Estonia in 2007 (Estonian MoD 2008:3). Three years later, in 2010, the government observed that the security environment was 'becoming increasingly difficult to predict, and are more versatile than ever before'. The impact of political confrontation, economic disputes, competition for resources, religious and ethnic tensions, failed states and non-state actors was, according to the government, often global. Globalisation was considered to bring along the rapid proliferation of security threats:

With the growth in global population, demand for food, energy and other resources is increasing. Tensions over natural resources are thus more probable. Conflicts, tensions and instability may also be caused by the reduction or unequal division of arable land, fresh water and other natural resources. Deterioration of the environment, especially climate change, may add to instability, as it is the poorest and most vulnerable areas that are often affected, and the probability of natural disasters is increasing.

(Estonian Government 2010:5)

Russia caused major concern for the Estonian government. Not least Russia's ambitions of restoring its status as a major global power and the Russian regime's willingness to contest other countries' preparedness to use military force to achieve its goals worried the Estonians. 'Russia also uses its energy resources as political and economic means in different areas of international relations', the government observed (Estonian Government 2010:7). 'A direct military attack against Estonia is unlikely; however, such a threat cannot be ruled out altogether', the government on the one hand concluded. On the other hand, the government observed that Russia had demonstrated an increased interest in re-establishing its spheres of influence and strengthening its influence over Europe's security environment. 'The presence of Russian Federation military forces in close proximity

84 *The empirical exploration*

to the Estonian border has increased', the government noted (Estonian MoD 2011:7). 'The international security environment is tense', the government concluded in 2017. 'Threats and risks that emerged in connection with globalisation still exist and their influence has somewhat sharpened. The number of conflicts has not decreased', the government observed. Especially the unresolved disputes and conflict zones in Europe and its vicinity worried the government. 'European security is affected by Russia's increased military activity and aggressive behaviour', the government warned. 'Russia is interested in restoring its position as a great power and for that purpose will not refrain from coming into a sharp opposition with the West and the Euro-Atlantic collective security system', the government predicted (Estonian Government 2017:3–4). Migration flows, ideological and religious extremism, the state of the global economy as well as the impact of climate change were other topics elaborated on by the government.

7.3 Ends

In 2004, the Estonian government declared that the goals of its security policy was 'to preserve Estonia's independence and sovereignty, territorial integrity, constitutional order, and public safety' (Estonian Government 2004:3). The government made clear that its approach to ensuring these national interests was 'based on the conviction that security serves to enforce human rights, fundamental freedoms and core human values'. According to the government, adherence 'to democratic principles enables the persistent development of the society. This strengthens a viable civil society and the will to defend Estonia, and advances Estonia's international standing and reputation' (Estonian Government 2010:3). Notably, in 2010, the end of 'preserving' Estonia's independence et cetera had been changed to 'safeguarding' the very same aspects as well as 'the existence of the state and its people'. In addition, Estonia's security policy was said aiming not only at providing 'the basis for sustainable development and welfare' but also at protecting 'constitutional values' as well as preventing threats and 'responding to them in a swift and flexible manner' (Estonian Government 2010:4). In addition, the functioning and unity of the EU and NATO as well as strong transatlantic cooperation were considered fundamental Estonian ends. To best protect the common European and transatlantic interests, the government emphasised the necessity of upholding a political dialogue between the EU and NATO and strived for 'intensified co-operation and co-ordination in the field of crisis management' between the two organisations (Estonian Government 2010:10).

The Estonian 'strategy addresses courses of action that are aimed at deterring any possible military attack against Estonia', the government declared in 2011. 'The main task of Estonia's national defence is to prevent a possible military attack against Estonia and, should the need arise, to provide for successful defence', the government made clear. The government declared that one of its key objectives was to link Estonia's security 'with the security of its allies as tightly as possible, exercising effective co-operation and strengthening Estonia's positive reputation in bilateral and regional co-operation formats' (Estonian MoD 2011:5

and 8–9). Notably, the Estonian defence industry policy of 2013 aimed not only at enhancing Estonia's capacity to defend itself but also at increasing the national wealth (Estonian Government 2013). The prevention and combating of the threat to the security of the state and the functioning of the state were at the core of the national defence act of 2015 (Estonian Government 2015). In 2017, the overarching objectives of the Estonian security policy was slightly changed. Consequently, the independence, sovereignty, territorial integrity and constitutional order of the state, as well as the survival of the people and the safety of the population, were to be secured rather than safeguarded. The government presented a list of objectives, some with direct impact on the defence policy such as 'strengthen the collective defence, enhance its efficiency and deterrence value, focusing on the defence of the Alliance's territory and consolidate the presence of combat-ready allied forces in Estonia and its vicinity'. The other objectives included enhancing 'the competitiveness of the Estonian state and enterprises' (Estonian Government 2017:7–8).

7.4 Means

In 2009, developing essential capabilities for initial self-defence was at the top of the agenda regarding the means of Estonian armed forces. Hence, it was considered a necessity to have the capacity to resist potential aggression until the arrival of allied forces. The government listed a number of prioritized operations for which the means were to be developed. The response to a sudden attack, the reception of allied forces by air, sea, or land, the support to allied actions in Estonia, and the defence of strategically important areas and objects were all on the list. So were air defence as well as mine clearance and mine hunting in Estonia's territorial waters. Regarding the air defence, completing the reconstruction of Ämari Air Base and equipping the air defence battalion with Mistral surface-to-air missiles as well as with ZU-23-2 anti-aircraft cannons were given priority. The high-readiness infantry brigade was to remain the main priority for the land forces. One of the brigade's infantry battalions was to remain professional while another infantry battalion and about a third of all support units of the brigade were to be 'formed and trained from conscripts during an annual training cycle'. The upgrading of anti-tank capabilities was considered important. In addition, mechanised units were to be developed within the brigade's framework. 'This will be achieved by acquiring either tanks or infantry fighting vehicles', the government announced (Estonian MoD 2009:9). New multirole high-speed patrol vessels were to constitute the bulk of the navy while multifunctional helicopters were at the top of the agenda for the Estonian air force. 'An adequate search and rescue (SAR) capability is one of the preconditions for potential basing and operating air policing assets in Estonia', the government explained (Estonian MoD 2009:10). Developing capabilities to provide an early warning as well as 'modern, deployable, mobile, sustainable rapid response units with the capability to ensure military defence of the whole territory as well as to participate in operations outside Estonia' was presented as a main objective of Estonia's military (Estonian Government 2010:14).

In 2013, the plans were slightly altered. The brigade was now mentioned as the 1st Infantry Brigade. It included the professional Scouts battalion, which was to receive modern IFVs, while the rest of the brigade's manoeuvre battalions were to be equipped with armoured personnel carriers (APCs). In addition, by 2022, the 2nd Infantry Brigade was to be operational. Regarding air operations, the existing air surveillance radars were to be modernised, and additional medium range radars were to be procured. The bulk of the navy was to consist of three modernised mine countermeasure vessels and a naval auxiliary vessel. A special operations unit had, in addition, been organised. In total, 3,200 conscripts were to be drafted annually (Estonian MoD 2013). The government also stressed its ambition on further developments in the area of military cyber defence capabilities. Not least the establishment of the NATO Cooperative Cyber Defence Centre of Excellence in Estonia was seen as an important step in this regard (Estonian MoD 2014). In 2017, the government stated that Estonian military defence 'is built in line with the principle of territorial defence. For this reason, Estonia will develop both mobile units and territorial defence units' (Estonian Government 2017:11). The Estonian MoD announced several measures to be implemented in order to 'being even better able to defend Estonia'. Hence, the wartime rapid response structure was to grow from 21,000 to 25,000 troops. Conscript service was to increase to 4,000 conscripts annually. The 1st Infantry Brigade was to adopt CV90 IFVs for one of its battalions while the other two battalions were to use APCs. From the spring of 2017, the brigade was also to include a NATO heavy armoured battalion. By 2026, the 2nd Brigade was to be operational as a motorized light infantry unit. The brigade was to include two additional battalions, one infantry and one artillery, over what was initially planned. Moreover, the territorial defence structure was to increase by about 1,000 soldiers organised in more than ten light infantry companies. A cyber command was to be established. The key procurement included at least 12 K9 Thunder self-propelled howitzers from South Korea and 44 CV90 IFVs from Sweden. 'At any one time, Ämari Air Base hosts at least four NATO Baltic Air Policing mission fighters', the government informed (Estonian MoD 2017:8).

The government continued to put emphasis on the cyber dimension: 'We will develop the capacity for cyber operations by continuing to develop the Defence Forces' Cyber Command further, developing cyber attack capability and promoting "cyber conscription" where IT can be chosen rather than infantry for

Table 7.1 Main military resources of Estonia.

	2000	*2010*	*2020*
Army brigade	0	1	2
MBT	0	0	0
ACV/APC/IFV	32	88	180
Combat aircraft	0	0	0
Principle surface combatant	3 patrol craft	1 corvette	0

Source: International Institute for Strategic Studies (2000, 2010, 2020).

completing one's compulsory military service' (Estonian MoD 2019a:17). The acquisition of night vision equipment and long-range anti-tank systems was at the top of the agenda when modernising the land forces. In addition, the logistics and combat engineering battalion of brigades were to be further developed. The navy was to acquire patrol boats and develop its sea surveillance capacity. A modernisation of mine hunting vessels was also decided. The air force was to receive an M-28 aircraft and air surveillance radar with identification-friend-or-foe capability (Estonian MoD 2019b).

7.5 Ways

Membership in NATO, the EU, and other international organisations, as well as successful bilateral and multilateral relations, made it possible, the government argued in 2004, not only to defend Estonia's interests but also to achieve the goals of Estonia's security policy. However, for ensuring Estonia's national security, the alliance with the US was described as being of 'primary importance'. Consequently, the government announced that Estonia was 'developing extensive and close cooperation' with the US 'in all spheres of major importance'. US military presence in Europe was the 'cornerstone of European security', the government declared (Estonian Government 2004:11). In addition, the trilateral cooperation among the Baltic States, the good cooperative relations with the Nordic countries, as well as the bilateral relations with Germany and Poland had, according to the Estonian government, strengthened Estonia's defence capability. Moreover, the government argued that:

> Estonia's international reputation as a partner in security cooperation is directly dependent upon the nation's readiness and ability to contribute to NATO and EU operations. Estonia is ready to participate in international operations within the framework of NATO and the EU, as well as of the UN and other international organisations, to ensure peace and security and solve crises, and to participate in other crisis management operations along with its NATO and EU allies [. . .]. Estonia will defend itself in any circumstance and against an enemy of any superiority.
>
> (Estonian Government 2004:12–13)

Consequently, Estonian military defence was based on two pillars: (1) initial self-defence and (2) collective defence with Estonia's NATO membership as the cornerstone. 'These two together will create adequate deterrence to prevent any potential aggression', the government argued. 'The defence of Estonian territory against any large-scale attack will be conducted as a NATO collective defence operation together with Allies', the government made clear (Estonian MoD 2008:3). In 2010, the government declared its intention to ensure Estonia's security through memberships in NATO and the EU, 'as well as close co-operation with its allies and other international partners'. The former organisation was still considered as serving as the 'cornerstone of European security and defence [. . .].

88 *The empirical exploration*

Estonia ensures credible deterrence and military defence through NATO's collective defence' (Estonian Government 2010:4). Participation in the crisis management operations of NATO and the EU 'as well as in NATO Response Force and EU battlegroups will remain a priority for Estonia', the government announced (Estonian Government 2010:12). The government stressed that Estonia's security was inseparable from that of its allies. The collective defence dimension articulated in both the North Atlantic Treaty and the Treaty of Lisbon were used to strengthen the argumentation:

> Estonia's political weight in NATO and CSDP [Common Security and Defence Policy (of the EU)] decision-making processes depends on our active participation in providing solutions to key NATO and CSDP challenges. Estonia's relative contribution to NATO and the EU must be active and visible, considering Estonia's small size.
>
> (Estonian MoD 2011:9)

On the one hand, the government declared its willingness to contribute to the international military operations led by NATO, the EU, the UN or a coalition of the willing. On the other hand, the government explicitly announced that Estonian defence planning incorporated paramilitary operations, guerrilla activities and resistance movements (Estonian MoD 2011). The intention of the Estonian strategy was, according to the government, to:

> prevent and pre-empt threats, as well as counter them quickly and flexibly, should the need arise. Estonia will defend itself in any case, no matter how overwhelming the opponent might be. If the state temporarily loses control over a part of its territory, Estonian citizens will engage in organised resistance in that area.
>
> (Estonian Government 2017:3)

7.6 Conclusions: Estonian strategy

Obviously, the Estonian government has pledged its willingness to contribute to several international actors efforts regarding global crisis management. Not only the EU but also the UN, OSCE, and *ad hoc* coalitions of the willing have been mentioned in this context. However, in the end, only two of these actors are considered able to provide the support Estonia so desperately is seeking: NATO and the US. Even if the EU is mentioned as crucial for Estonia's security, this is rather mentioned in the context of political, economic and/or social security than in the context of national defence. Arguably, the Estonian government does not go so far in its supportive attitude to its key beneficiaries that bandwagoning can be considered a relevant label of the alignment strategy. We rather find it reasonable to define the balanced approach to both NATO and the US as *multiple-courting*. Since neither the EU nor the UN is completely out of the picture, at least not when it comes to other aspects of security than military as well as out-of-area military conflict management, we find this label most appropriate.

Clearly, the core ends of the military strategy are about *survival*. The Estonian government seldom, if ever, present any sincere elaborations on gaining international status and/or influence. We also consider it reasonable to conclude that the means of the strategy are designed for *national defence* rather than for expeditionary warfare. The challenge for the government has rather been on balancing the means essential for fighting the initial phase of a potential armed aggression solely on its own with the means necessary to receive and host allied reinforcements. The Estonian government is obviously aware of Estonia's geopolitical location as well as of its limited resources. Consequently, a *multilateral approach* to the ways is a necessity rather than an option. The government is nevertheless communicating its willingness to fight to the end for the survival of the Estonian sovereignty, even with guerrilla warfare if necessary.

Table 7.2 Estonian strategy.

Alignment strategy	Military strategy		
	Ends	Means	Ways
Multiple-courting	Survival	National defence	Multilateral approach

Already in 2000, Estonian President Lennart Meri clarified that Estonia was striving for dual memberships in both the EU and NATO. It is interesting to note that he mentioned the membership in the World Trade Organization (WTO) alongside the other two. We find his statement that Estonia, in non–Article 5 matters, moved towards developing a specific European capability, supporting our conclusions regarding alignment strategy (Meri 2000; see also Kuus 2003; Kasekamp 2020). Consequently, we do not fully agree with Toomas Riim when arguing that Estonia sought NATO membership as its primary goal for its foreign and security policy (Riim 2006). We rather tend to agree with Viljar Veebel and Illimar Ploom when arguing that, while Estonia's main security concerns are related to the ability to deter Russia, security reforms have also been considered necessary in terms of Estonian competitiveness and long-term economic sustainability (Veebel and Ploom 2016).

Consequently, we find the conclusions of David Takacs supporting our position in this regard. Collective defence treaties and resilience capabilities have both been considered necessary when deterring Russian aggression, and Estonia has continuously focused on improving their military and non-military deterrence measures together with NATO and the EU, respectively (Takacs 2017; see also Scott 2018; Veebel 2019). In addition, we also find Živilė Marija Vaicekauskaitė supporting our conclusion when arguing that Estonia has paid particular attention to bilateral defence cooperation with the US, 'who has always been seen as the key ally in shoring up their national defence and deterrence posture' (Vaicekauskaitė 2017:14; see also Praks 2019; Studemeyer 2019).

We agree with Holger Mölder when observing that Estonia's political and military elites took a positive attitude towards international cooperation and combat experience achieved through participation in international military missions. However, we do not support the idea that these missions had any crucial impact on the future practices, tactics and doctrines of the Estonian armed forces. Contrary, we argue that the elites focused on national defence rather than on expeditionary warfare (Mölder 2014; see also Mölder 2013). Consequently, we agree with Valdis Otzulis and Žaneta Ozoliņa when observing the challenges involved in the defence budgets and balancing the needs related to HNS with the needs associated with developing national capacities crucial for the initial phase of an armed aggression. We fully agree with the conclusion that affording the necessary developments of military points of entry like seaports and airports for force reception is fundamental for a small state sharing a border with Russia (Otzulis and Ozoliņa 2017; see also Odehnal et al. 2020). At the same time, we agree with Māris Andžāns and Viljar Veebel when they shed light on the other fundamental challenge, that is finding the balance between external military solidarity on the one hand and territorial defence on the other (Andžāns and Veebel 2017). We find the observations of Viljar Veebel and Illimar Ploom interesting in this regard. If they are right, Estonian 'military tends to interpret resilience along the logic of total defence, while in the civilian sense resilience stands rather for a well-functioning society, a strong economy and widely shared well-being' (Veebel and Ploom 2018:20). Arguably, the cyber domain overlaps these positions (see, for example, Cardash et al. 2013; Crandall and Allan 2015; Molis et al. 2018).

Notes

1 We would like to express our gratitude to Colonel Mirko Arroküll, Embassy of Estonia, Stockholm and Lieutenant Colonel Jörgen Westerlund, Embassy of Sweden, Tallinn for their support.
2 Mart Laar of the Pro Patria Union served as prime minister 25 March 1999–28 January 2002. Siim Kallas of the Reform Party (ERE) served as prime minister 28 January 2002–10 April 2003. Juhan Parts of the Res Publica Party (RP) served as prime minister 10 April 2003–12 April 2005. Andrus Ansip of the ERE served as prime minister 12 April 2005–26 March 2014. Taavi Rõivas of the ERE served as prime minister 26 March 2014–23 November 2016. Jüri Ratas of the Centre Party (EKE) served as prime minister 23 November 2016–26 January 2021. As of 1 January 2022, Kajsa Kallas of the ERE had served as prime minister since 26 January 2021. See www.valitsus.ee/en.

Bibliography

Andžāns, Māris and Viljar Veebel (2017). 'Deterrence Dilemma in Latvia and Estonia: Finding the Balance between External Military Solidarity and Territorial Defence' *Journal on Baltic Security* Volume 3, Issue 2.

Cardash, Sharon, Frank Cilluffo and Rain Ottis. (2013). 'Estonia's Cyber Defence League: A Model for the United States?' *Studies in Conflict & Terrorism* Volume 36, Issue 9.

Crandall, Matthew and Collin Allan (2015). 'Small States and Big Ideas: Estonia's Battle for Cybersecurity Norms' *Contemporary Security Policy* Volume 36, Issue 2.

Estonian Government (2004). *National Security Concept.*
——— (2010). *National Security Concept.*
——— (2013). *Defence Industry Policy.*
——— (2015). *National Defence Act.*
——— (2017). *National Security Concept.*
Estonian Ministry of Defence (MoD) (2008). *Cyber Security Strategy.*
——— (2009). *Long Term Defence Development Plan 2009–2018.*
——— (2011). *National Defence Strategy.*
——— (2013). *National Defence Development Plan 2013–2022.*
——— (2014). *Cyber Security Strategy.*
——— (2017). *National Defence Development Plan 2017–2026.*
——— (2019a). *Cyber Security Strategy.*
——— (2019b) *Defence Action Plan 2020–2023.*
International Institute for Strategic Studies (IISS) (2000). *The Military Balance 2000–2001.* Oxford: Oxford University Press.
——— (2010). *The Military Balance 2010.* London: Routledge.
——— (2020). *The Military Balance 2020.* London: Routledge.
Kasekamp, Andres (2010). *A History of the Baltic States.* Basingstoke: Palgrave Macmillan.
——— (2020). 'An Uncertain Journey to the Promised Land: The Baltic States' Road to NATO Membership' *Journal of Strategic Studies* Volume 43, Issue 6–7.
Kuus, Merje (2003). 'Security in Flux: International Integration and the Transformations of Threat in Estonia' *Demokratizatsiya* Volume 11, Issue 4.
Liddell-Hart, Basil (1997). *History of the Second World War.* London: Papermac.
Meri, Lennart. (2000). 'Estonia's Security and Defence Policy New Steps towards NATO Membership' *The RUSI Journal* Volume 145, Issue 3.
Molis, Arūnas, Claudia Palazzo and Kaja Ainsalu (2018). 'Mitigating Risks of Hybrid War: Search for an Effective Energy Strategy in The Baltic States' *Journal on Baltic Security* Volume 4, Issue 2.
Mölder, Holger (2013). 'The Development of Military Cultures' in Tony Lawrence and Tomas Jermalavičius (eds). *Apprenticeship, Partnership, Membership: Twenty Years of Defence Development in the Baltic States.* Tallinn: International Centre for Defence Studies.
——— (2014). 'Estonia and the ISAF: Lessons Learned and Future Prospects' *The Polish Quarterly of International Affairs* Volume 23, Issue 2.
Odehnal, Jakub, Jiří Neubauer, Lukáš Dyčka and Tereza Ambler (2020). 'Development of Military Spending Determinants in Baltic Countries: Empirical Analysis' *Economies* Volume 8, Issue 3.
Otzulis, Valdis and Žaneta Ozoliņa (2017). 'Shaping Baltic States Defence Strategy: Host Nation Support' *Lithuanian Annual Strategic Review* Volume 15, Issue 1.
Praks, Hendrik (2019). 'Estonia's Approach to Deterrence: Combining Central and Extended Deterrence' in Nora Vanaga and Toms Rostoks (eds). *Deterring Russia in Europe: Defence Strategies for Neighbouring States.* Abingdon: Routledge.
Riim, Toomas (2006). 'Estonia and NATO: A Constructivist View on a National Interest and Alliance Behaviour' *Baltic Security & Defence Review* Volume 8, Issue January.
Scott, David (2018). 'China and the Baltic States: Strategic Challenges and Security Dilemmas for Lithuania, Latvia and Estonia' *Journal on Baltic Security* Volume 4, Issue 1.
Studemeyer, Catherine Cottrell. (2019). 'Cooperative Agendas and the Power of the Periphery: the US, Estonia, and NATO after the Ukraine Crisis' *Geopolitics* Volume 24, Issue 4.

Takacs, David (2017). 'Ukraine's Deterrence Failure: Lessons for the Baltic State' *Journal on Baltic Security* Volume 3, Issue 1.

Taylor, Neil (2018). *Estonia: A Modern History*. London: C. Hurst & Co.

Vaicekauskaitė, Živilė Marija (2017). 'Security Strategies of Small States in a Changing World' *Journal on Baltic Security* Volume 3. Issue 2.

Veebel, Viljar (2019). 'Researching Baltic Security Challenges after the Annexation of Crimea' *Journal on Baltic Security* Volume 5, Issue 1.

Veebel, Viljar and Illimar Ploom (2016). 'Estonian Perceptions of Security: Not Only about Russia and the Refugees' *Journal on Baltic Security* Volume 2, Issue 2.

―――― (2018). 'Estonia's Comprehensive Approach to National Defence: Origins and Dilemmas' *Journal on Baltic Security* Volume 4, Issue 2.

8 The strategy of Hungary[1]

The primary sources analysed in this chapter consist of the basic principles of the security and defence policy adopted by the Hungarian NA in 1998, the NSSs of 2004, 2012 and 2020, the military strategy of 2012, as well as the NCSS of 2013.[2] Since these official Hungarian documents do not explicitly present the organisation and number of units, we have decided to complement the empirical material with the information regarding means and ways presented by the International Institute for Strategic Studies (IISS).

8.1 Historical background

The Kingdom of Hungary was established about 1000 when Stephen I was crown as the first king and lasted until 1526 when King Louis II fell at the Battle of Mohács against the invading Ottomans. During its five centuries of existence, the Kingdom waged several wars against Austria, Bavaria, Bohemia, Bulgaria, Byzantine, Croatia, the Holy Roman Empire, Lithuania, Milan, Moldavia, the Mongols, Naples, the Ottomans, Poland, Serbia, Venice and Wallachia. Following the Hungarian defeat at the Battle of Mohács, Hungary was portioned between the Ottoman Empire, the Habsburg Monarchy, and the Principality of Transylvania. Moreover, the death of Louis II also marked the end of the Jagiellonian dynasty, passing the dynastic claims to the House of Habsburg. The Austro-Hungarian Compromise of 1867 partially re-established the sovereignty of the Kingdom of Hungary as a part of the dual monarchy of Austria-Hungary. Following its defeat in WWI, Austria-Hungary was dissolved in 1918. Soon afterwards, neighbouring Czechoslovakia, Romania and Serbia all annexed pats of the Hungarian kingdom. In June 1920, the Treaty of Trianon was signed, formally ending WWI for Hungary. After the signing, the remaining parts of Hungary formed a new state and kept its status as kingdom with Admiral Miklós Horthy as acting regent. Between 1938 and 1940, Hungary regained, through the arbitration of Nazi-Germany and Italy as well as by brute force, territories lost to Czechoslovakia and Romania following the Trianon Treaty. In late 1940, Hungary formally joined the Axis alliance. In 1941, Hungary took part in the Axis invasions of Yugoslavia, hence regaining some territory lost to Serbia, and of the USSR. The Hungarian attempt to leave the Axis alliance led to German occupation in 1944 and eventually an invasion and

DOI: 10.4324/9781003298052-10

94 *The empirical exploration*

occupation by the USSR. Following the Paris Treaty in 1947, Hungary's territory was restored to the borders of the Trianon Treaty. In 1956, following public unrest and uprising, Hungary was once again invaded by the Red Army, and the communist regime was reinstalled (see, for example, Liddell-Hart 1997; Cartledge 2011; Hill 2004).

8.2 Strategic environment

In 1998, the Hungarian government concluded that, compared to era of the bipolar world order, 'the threats of a worldwide armed conflict has decreased to a minimum while at the same time, the circle of risks and sources of threat has significantly increased and become more complex'. Differing levels of social development among countries, economic and financial crisis, ethnic and religious tensions, terrorism, organised crime, trafficking, proliferation of WMD, mass migration and environmental hazards, as well as attacks against information systems, were all elaborated on. A particular feature of security was, the government argued, that new and traditional 'challenges often emerge simultaneously and reinforce each other' (Hungarian NA 1998:2). Cross-border threats such as the spreading of the technological skills required for the production and potential use of WMD including their means of delivery were in focus of the considerations in 2004. Unstable regions, failed states, illegal migration, and economic instability were, together with the challenges of the information society, also elaborated on in general terms. The situation in the Caucasian, the Central Asian and the Mediterranean regions as well as the Middle East was also analysed. The government gave specific attention to the developments in the Commonwealth of Independent States (CIS). 'Russia continues to be an important factor in international politics because of its geographical scope, its natural and human resources and its military potential, and its nuclear power in particular. The dangers emanating from the country's internal instability have decreased but not yet completely disappeared', the government observed (Hungarian MoFA 2004:8). In 2012, János Martonyi, the Hungarian Minister for Foreign Affairs, declared that in:

> a historical context, today's Hungary enjoys an unprecedented level of security. This has come about as a result of the successful process of Euro-Atlantic integration, solidarity and cooperation between Hungary and its Allies and their joint action against foreign and global threats.
> (Hungarian MoFA 2012:2)

The government concluded that the level of the threat of a conventional attack against Hungary and its allies was marginal. At the same time, the government observed that not all conflicts of the past decades had been resolved. The security in 'certain neighbouring regions' was perceived as fragile. The government also notified the intensive build-up of modern military capabilities in some of these regions. Notably, the government used the phrase 'close geographical proximity to Hungary' in its elaborations without always defining exactly which areas the analysis were referring to. However, in some parts of the considerations, Hungary's eastern and southern neighbourhoods were given specific attention. 'Given the presence of Hungarian communities in the neighbouring countries, the particularities of their situation and their relationship with the mother country, the security of these countries is inseparable from Hungary's own security and vice versa', the government argued (Hungarian MoFA 2012:8).

Poverty, instability, lack of democracy, extremism, terrorism, global climate and environmental change, along with trafficking in arms, drugs and/or humans, were considered as constituting hotbeds for violence within countries and regions. Considering Hungary's dependency on energy import, regions important for the country's energy requirements were given specific attention. 'Stable and predictable energy supply available at a competitive price is of strategic importance', the government admitted (Hungarian MoFA 2012:14). 'Hungary regards the Central and Eastern European region with special attention, where cybersecurity can be further improved within the framework of regional cooperations', the government announced when presenting its NCSS (Hungarian NA 2013:4). In 2020, the government concluded that the global security environment had undergone fundamental changes during the previous decade. 'The new challenges are based on the emerging, multipolar world order', the government concluded as it also paid attention to accelerating climate and demographic change, mass migration, the depletion of natural resources and the effects of the technological revolution. 'The competition for power is increasingly extending to global public goods: there is an increasing struggle for control of international waters and their resources, the Arctic and outer space, and the dominance of cyberspace', the government observed (Hungarian MoFA 2020:1 and 8–9).

8.3 Ends

The overarching ends of the Hungarian security policy were, the government declared in 1998:

- to guarantee the independence, sovereign statehood and territorial integrity of the country [. . .];
- to promote the practice of the rule of law and the undisturbed functioning of the democratic institutions and the market economy and to contribute to the internal stability of the country;
- to promote the full implementation of citizens', human as well as national and ethnic minority rights [. . .];
- to create the necessary conditions for the safeguarding of the safety as well as material and social security of all people living on the territory of the Republic [. . .] and for the safeguarding of national wealth;
- to contribute to the implementation of the contents of the North Atlantic Treaty and to the security of Allies;
- to contribute to the maintenance of international peace and to the enhancement of security and stability in the Euro-Atlantic region, in Europe and its immediate neighbourhood;

96 *The empirical exploration*

- to contribute to ensuring the conditions for the international economic, political, cultural and all other kinds of relations and co-operation of the Republic of Hungary.

(Hungarian NA 1998:2)

The government was also 'paying distinguished attention to the situation of the Hungarian communities living beyond' Hungary's borders and 'to the implementation of their interests in harmony with the norms of international law' (Hungarian NA 1998:3). The preservation of the cultural heritage and identity of the Hungarian people was also mentioned as a central end in the NSS of 2004. Notably, forms of self-government and autonomy for Hungarians living in neighbouring countries was declared a goal of the government. The implementation of political pluralism and freedom of enterprise, the preservation of the constitutional order as well as Hungary's 'rapprochement to the level of the developed industrial nations' and ensuring a sustainable economic growth were also mentioned as national security interests (Hungarian MoFA 2004:2). Guaranteeing Hungary's and its citizens' security as well as citizens' freedom and welfare was central also in the security strategy of 2012. Notably, the strategy also mentioned the security of the Hungarian nation as well as Hungary being a confident and strong player on the international stage as ends. 'The preservation of the engagement of the United States in Europe and the strengthening of its strategic partnership with the European Union is in Hungary's interest', the government declared (Hungarian MoFA 2012:7). The government also announced its interest in creating a defence capability against ballistic missiles at the NATO level. The performance of Hungarian military in international operations 'has a significant impact on Hungary's international prestige and influence, contributing to the increase of the country's ability to assert its interests', the government argued (Hungarian MoD 2012:6).

'Fulfilling international obligations contributes to the strengthening of international peace and security and, consequently, Hungary's security, increasing the international prestige of the country at the same time' the government claimed (Hungarian MoD 2012:17). 'The protection of Hungary's sovereignty in the Hungarian cyberspace is a national interest', the government announced in 2013 as it presented Hungary's NCSS (Hungarian NA 2013:3). The 'preservation of our Hungarian mother tongue and culture beyond our borders' was announced as a national interest in 2020. To increase Hungary's international competitiveness, to develop the industrial capacities in the defence sector and to improve Hungary's demographic situation were other interests articulated by the government. According to the government, the key to the survival of the Hungarians was a strong, nationally based Hungary. The country's 'millennial statehood, our Hungarian language and culture, our history and traditions, our Christian values' were hence fundamental, the government argued. 'Our primary security policy interest is to protect, preserve and strengthen the Hungarian state's self-determination and freedom of action in the constantly changing conditions', the government declared (Hungarian MoFA 2020:2–3). The government also stressed Hungary's status as a member of NATO and the EU.

8.4 Means

In 1998, the government announced its ambition maintaining and developing Hungary's armed forces in order 'to avert an armed attack'. The government stressed its awareness of the requirements for continuously having a credible defence capability based on the full range of capacities. The focus was on defending Hungary and on contributing to the defence of NATO allies. The government also declared its willingness to put Hungarian armed forces at the disposal of 'other missions of the Alliance in accordance with its possibilities' (Hungarian NA 1998:5). According to the IISS, in 2000 the bulk of the Hungarian land forces consisted of two mechanised divisions with, altogether, seven mechanised brigades and over 1,000 MBTs. Most of the MBTs were rather outdated T-55 and only 238 more modern T-72 models. In addition, almost 500 BMP-1 IFVs and a similar number of BTR-80 APCs were at their disposal. Although the Hungarian air force had almost 70 combat aircraft, only 12 MiG-29A were available. All MiG-21 were to be taken out of service by 2001. In addition, the air force flew 24 Mi-24 attack helicopters (IISS 2000). Some years later, the approach of the government had another focus. 'Hungary needs to possess the military capabilities required for the collective defence of Hungary within the framework of NATO and for the collective defence of its allies, as well as capabilities required for crisis management and peacekeeping operations and disaster-relief launched with the participation of NATO-allies', the government declared (Hungarian MoFA 2004:11). Rapidly deployable and sustainable forces suited for expeditionary operations and able to cooperate with allied forces were hence considered a key for success. Moreover, these forces were to be deployed and used without any geographical limitations. The government admitted that the development of the necessary capabilities had to take place in close cooperation within the NATO and EU framework 'and by making use of the opportunities lying in bi- and multilateral international cooperation and development programs'. Notably, the government not only pledged developing and procuring modern equipment but also developing 'an armed force that is new in the sense of operational philosophy' (Hungarian MoFA 2004:21–22).

In 2005, the Hungarian army organised two infantry brigades with altogether seven light infantry battalions, one armoured battalion and two reconnaissance battalions. The main equipment of the land forces consisted of 238 T-72 MBTs and 178 BTR-80 IFVs. Four additional reserve brigades were organised. The air force organised one tactical fighter wing with 14 MiG-29 Fulcrum and one attack helicopter squadron with twelve Mi-24 Hind helicopters (IISS 2005). In 2012, the government pledged to develop 'the required capabilities in both national and collective frameworks' (Hungarian MoFA 2012:7). 'The Hungarian Defence Forces are on the road to modernization', Hungarian Minister of Defence Csaba Hende announced. 'We have to abandon the previously denizened, comfortable, but altogether dangerous attitude that national defence is feasible without substantive military strength, relying on a bare minimum of own capabilities, and trusting solely in the solidarity of NATO and EU nations' (Hungarian MoD 2012:1). The voluntary reserve system was declared an integral part of the armed force,

98 *The empirical exploration*

Table 8.1 Main military resources of Hungary.

	2000	2010	2020
Army brigade	4	2	2
MBT	806	30	44
ACV/APC/IFV	910	328	392
Combat aircraft	68	27	14

Source: International Institute for Strategic Studies (2000, 2010, 2020).

of which regular parts consisted of professional and contracted personnel. The government stressed the importance of Hungarian participation in international military operations even in strategic distances. 'Hungary, based upon its concerning level of ambition, is ready to and capable of stationing one thousand troops in international operations at any given moment', the government pledged (Hungarian MoD 2012:11). According to the IISS, in 2013 the Hungarian army organised two brigades in a total six manoeuvre battalions and two logistic battalions. In addition, one special operations battalion was organised. The army operated 30 T-72 MBTs. The air force had taken all MiG combat aircraft out of service. Instead, 14 JAS39 Gripen combat aircraft were leased from Sweden. In addition, the air force flew 11 Mi-24 attack helicopters (IISS 2013).

In 2015, the government announced that the introduction of the voluntary reserve system was an ongoing process. The system consisted of 8,000 reservists (Permanent Mission of Hungary to the OSCE 2015). According to the IISS, in 2018 the Hungarian army organised two brigades; the 5th Mechanised Infantry Brigade with one armoured reconnaissance battalion and three light infantry battalions, and the 25th Mechanised Infantry brigade with one tank battalion and two light infantry battalions. In addition, one special forces regiment was organised. The air force was organised in one squadron with 14 JAS39 Gripen combat aircraft and one squadron of 11 Mi-24 Hind attack helicopters. The IISS also announced the Hungarian contributions to the EUBG organised in cooperation with the four Visegrád countries. This unit was to be on standby for a second term in 2019. Moreover, the IISS referred to the Zrínyi 2026 plan for the modernisation of the armed forces. According to the IISS, announced modernisations gave priority to individual soldier equipment and to fixed, as well as rotary-wing, aircraft (IISS 2018). 'In the two decades following the change of regime, no comprehensive, system-level development took place' in the armed forces, 'and before 2010 there was a deliberate downsizing' of the military, 'which endangered the security of Hungary', the government admitted in 2020. The 'current security challenges and modern warfare requires complex and costly weapon systems, which necessitates the development of the domestic defence industry', the government concluded (Hungarian MoFA 2020:5).

8.5 Ways

In 1998, the government announced that it based its security policy on two fundamental pillars: national self-strength on the one hand, and Euro-Atlantic integration as well as international cooperation on the other. Not only NATO but the Council of Europe, the EU, the Organization for Economic Co-operation and Development (OECD), OSCE and the UN as well were mentioned in this regard. In providing for its security, Hungary 'attaches particular importance also to the regional and sub-regional as well as the traditional bilateral forms of international co-operation' (Hungarian NA 1998:3). The bulk of the Hungarian contributions to international military operations in 2000 consisted of an engineer battalion to the NATO-led SFOR (Stabilisation Force) in Croatia, and some 100 UN peacekeepers to UNFICYP (United Nations Peacekeeping Force in Cyprus) (IISS 2000). The effectiveness of the Common Foreign and Security Policy (CFSP) and the ESDP of the EU as well as the transatlantic cohesion, including the US military presence and active engagement in Europe, were all central in the elaborations on the ways. NATO was perceived as essential for the transatlantic security, and the government declared that Hungary was

> striving for a substantial contribution to the entire [spectrum] of allied missions, including participation in expeditionary operations outside the Euro-Atlantic region and ensures for the Hungarian defence forces to be able to contribute in a timely manner, and with appropriately trained and prepared forces to NATO-led operations, as well as to coalition-type operations carried out with allied participation.
>
> (Hungarian MoFA 2004:11)

In 2005, the Hungarian had one mechanised battalion deployed to the NATO-led KFOR in Kosovo and some 130 troops deployed to ISAF in Afghanistan. In addition, an engineer unit with some 150 troops were deployed to the EU-led EUFOR (European Union Force) Althea in Bosnia-Herzegovina, while some 84 peacekeepers were deployed to UNFICYP. Active contribution to NATO's collective defence and security was described as Hungary's most important security policy obligation in 2012. 'Hungary wishes to continue to take an active part in NATO and EU-led crisis management activities, operations and missions', the government announced (Hungarian MoFA 2012:7). Preventive and protective measures were hence given priority. At the same time, the government declared that the core task of the Hungarian armed forces was 'the armed defence of Hungary's independence, territory, airspace, population and material goods against any external attack either individually or in an Allied framework' (Hungarian MoD 2012:17). Consequently, providing HNS for allied forces was also mentioned as a core task. The IISS reported that in 2012, Hungary had one light infantry company deployed each to ISAF, KFOR and EUFOR Althea. In addition, one infantry platoon served with UNFICYP (IISS 2013).

These deployments remained more or less the same in 2018. However, the contributions to ISAF had been replaced with participation in the NATO-led Operation Resolute Support in the same country. In addition, some 140 troops participated in the US-led Operation Inherent Resolve against the Islamic State

100 *The empirical exploration*

in Iraq (IISS 2018). 'It is in Hungary's interest to participate in or play a leading role in regional multinational formations and capability development initiatives', the government declared in 2020 (Hungarian MoFA 2020:16). The cooperation within the framework of Visegrád Cooperation and the bilateral defence relations with Germany, Poland, and the US were given priority. France, Italy and Turkey were also mentioned in this context.

8.6 Conclusions: Hungarian strategy

Notably, the Hungarian government included not just actors ready to assume responsibility to lead international military operations in its elaborations on relevant strategic partners. The government rather took a broad approach to security and included other organisations such as the OSCE as well. Arguably, this indicates a hedging alignment strategy based on *multiple-courting*. Clearly, the Hungarian government struggled with economic constraints, which made the OECD and especially the EU fundamental regarding the political, the economic and/or the social aspects of security. The government also presented rather broad considerations regarding the ends. On the one hand, the government seemed determined to defend the *survival* of the Hungarian nation rather than the Republic of Hungary. The challenges caused by increased and often uncontrolled migration following the situation in the Middle East were often related to demographic concerns. Notably, the armed forces had been used to strengthening the control of Hungarian borders. The government also placed specific attention on the situation of the ethnic Hungarians living outside Hungary. On the other hand, the government put pride in the traditions of the Hungarian state and emphasised the need for gaining international recognition. *Status* rather than influence was hence considered important in the international relations. Therefore, we find it reasonable to argue that the end of the Hungarian strategy consists of a blend of survival and status.

We find it reasonable to conclude that the means of the strategy mainly has been designed for *national defence*. Despite the explicit pledges of the government to contribute to the international efforts of its allies, the development of the Hungarian armed forces has been less impressive. Notably, the government seldom refers to numbers when it comes to the means. The contributions have often been companies rather than in battalions, and the companies have been deployed to separate operations led by each of the main actors rather than concentrated in one mission. Consequently, this approach in itself supports our claim on multiple-courting. Notably, three decades after the end of the Cold War, the means of the army still have a Soviet touch. Finally, regarding the ways, a *multilateral approach* may very well have been considered a necessity. With outdated MBTs, IFVs and APCs for the army, as well as personal equipment for the individual soldiers, and having to lease rather than procure combat aircraft for the air force, the capacity of the armed forces simply provides no option. However, the tendencies towards increased emphasis on unilateralism may lead to a shift in the Hungarian strategy in this regard in the near future.

Table 8.2 Hungarian strategy.

Alignment strategy	Military strategy		
	Ends	Means	Ways
Multiple-courting	Survival/status	National defence	Multilateral approach

Obviously, we do not agree with Mark Yaniszewski when arguing that Hungary had made 'tremendous progress' since the fall of communism. Rather we agree with his observation that the financial constraints have delayed the necessary reforms of the military (Yaniszewski 2002). We also find the observations of Mária Vass supporting our conclusion regarding an alignment strategy based on multiple-courting (Vass 2002). We agree with Pál Dunay when arguing that the Hungarian armed forces have continuously contributed to international military operations despite the lack of relevant transformation of the armed forces. We also agree with Dunay when concluding that the 'shortcomings of the Hungarian military had to keep its participation overall limited' in this regard, although demonstrating 'loyalty towards its main western supporters' (Dunay 2005a:52). We support Dunay when arguing that 'Hungary's defence reform performance since it has joined NATO has primarily generated disappointment amongst its allies. The United States in particular has been vocal in its criticism'. Moreover, we tend to agree with him when claiming that 'Hungary has managed to compensate for its weak performance in defence reform through the promises it has made. It has been extremely skilful in making promises and seldom delivering on them later' (Dunay 2005b:29–30; see also Biro 2005).

We agree with Gábor Zord when shedding light on the difficulties of not only Hungary but of most former Warsaw Pact members in the transiting away from the legacy of Soviet standards (Zord 2007, see also Korom 2019). Gergely Varg's illumination of the slow and painful transformation of defence in Hungary is arguably also proving support to our conclusions, not least regarding the focus on national defence (Varga 2011). We find the observations of Imre Takács regarding the experience of the Hungarian armed forces to 'deploy its capabilities in domestic matters' interesting (Takács 2005:19). We also find the claims of Joseph Bell and Ryan Hendrickson noteworthy. When arguing that Hungary has taken a bystander position within NATO which, in the long-run, may impact negatively on its quest for recognition among its allies, they conclude that this approach risks decreasing Hungary's 'political relevance by feeding into the free-rider reputation' (Bell and Hendrickson 2012:160; see also Csizmazia 2019). We do not go as far as Zsuzsanna Végh when accusing Hungary of being a Russian Trojan Horse within the Western community. However, we do agree that the Hungarian dependency on the energy of and its closer relations with Russia due to these matters may have weakened Hungary's position within the EU as well (Végh 2015). As Juraj Marušiak observes, this dependency and the Hungarian policy towards Russia may also weaken the cohesion among the Visegrád counties (Marušiak 2015;

see also Lazar 2019). On the one hand, these findings also weaken our argument regarding multiple-courting and hence mainly towards the EU and NATO. On the other hand, the multiple-courting of the Hungarian government tend not to exclude even Russia. It is, however, too early to conclude whether this indicates a shift towards a balance of power strategy. Consequently, we agree with Tamás Csiki when arguing that the fragmented attempts to transform the Hungarian armed forces on the one hand have left the armed forces with outdated equipment and, on the other hand, have exploited the misunderstanding among Hungary's political elite what true multinational defence cooperation really means (Csiki 2015). Notably, this is in line with Bence Nemeth when arguing that the strategic documents presented by the government have been 'too general for providing appropriate input for defense planning' (Nemeth 2016:333).

Notes

1 We would like to express our gratitude to Colonel Andrea Gombos, Embassy of Hungary, Stockholm, and Lieutenant Colonel Rickhard Nordfjäll, Embassy of Sweden, Budapest, for their support.
2 Viktor Orbán of Fidesz (Hungarian Civic Alliance) served as prime minister 6 July 1998–27 May 2002. Péter Medgyessy served as independent prime minister 27 May 2002–29 September 2004. Ferenc Gyurcsány of the Socialist Party (MSZP) served as prime minister 29 September 2004–14 April 2009. Gordon Bajnai served as independent prime minister 14 April 2009–29 May 2010. As of 1 January 2022, Viktor Orbán of Fidesz had served as prime minister since 29 May 2010. See www.kormany.hu/en.

Bibliography

Bell, Joseph and Ryan Hendrickson (2012). 'NATO's Visegrad Allies and the Bombing of Qaddafi: The Consequence of Alliance Free-Riders' *Journal of Slavic Military Studies* Volume 25, Issue 2.

Biro, Istvan (2005). *The National Security Strategy and Transformation of the Hungarian Defense Forces*. Carlisle, PA: U.S. Army War College.

Cartledge, Bryan (2011). *The Will to Survive: A History of Hungary*. New York: Columbia University Press.

Csiki, Tamás (2015). 'Hungary's 15 Years within NATO: Lessons Learnt and Unlearnt' in Robert Czulda and Marek Madej (eds). *Newcomers No More? Contemporary NATO and the Future of the Enlargement from the Perspective of "Post-Cold War" Members*. Warsaw: International Relations Research Institute.

Csizmazia, Gábor (2019). 'Relations between the United States and Hungary: Phases and Fluctuations of the Last Two Decades' in Anna Péczeli (ed). *The Relations of Central European Countries with the United States*. Budapest: Dialóg Campus.

Dunay, Pál (2005a). 'Peace Operations and Humanitarian Assistance: The Contribution of Hungary' in Timothy Edmunds and Marjan Malesic (eds). *Defence Transformation in Europe: Evolving Military Roles*. Amsterdam: IOS Press.

——— (2005b). 'The Half-Hearted Transformation of the Hungarian Military' *European Security* Volume 14, Issue 1.

Hill, Raymond (2004). *Hungary*. New York: Facts On File Inc.

Hungarian Ministry of Defence (MoD) (2012). *National Military Strategy*.

Hungarian Ministry of Foreign Affairs (MoFA) (2004). *National Security Strategy*.
——— (2012). *National Security Strategy*.
——— (2020) *National Security Strategy*.
Hungarian National Assembly (NA) (1998). *Basic Principles of the Hungarian Security and Defence Policy*.
——— (2013). *National Cyber Security Strategy of Hungary*.
International Institute for Strategic Studies (IISS) (2000). *The Military Balance 2000–2001*. Oxford: Oxford University Press.
——— (2005). *The Military Balance 2005–2006*. Abingdon: Routledge.
——— (2010). *The Military Balance 2010*. London: Routledge.
——— (2013). *The Military Balance 2013*. Abingdon: Routledge.
——— (2018). *The Military Balance 2018*. Abingdon: Routledge.
——— (2020). *The Military Balance 2020*. London: Routledge.
Korom, Ferenc (2019). 'Hungarian Defence Forces Capability Transformation: Balancing Acquisition and Innovation' *Hungarian Defence Review* Volume 147, Issue 1–2.
Lazar, Zsolt (2019). 'Success and Failures of the Gripen Offsets in the Visegrad Group Countries' *Defense & Security Analysis* Volume 35, Issue 3.
Liddell-Hart, Basil (1997). *History of the Second World War*. London: Papermac.
Marušiak, Juraj (2015). 'Russia and the Visegrad Group: More Than a Foreign Policy Issue' *International Issues & Slovak Foreign Policy Affairs* Volume 24, Issue 1–2.
Nemeth, Bence (2016). 'Making Defense Planning Meaningful: The Evolution of Strategic Guidance in the Hungarian Ministry of Defense' *Defense & Security Analysis* Volume 32, Issue 4.
Permanent Mission of Hungary to the OSCE (2015). *Note Verbale*, 16 April.
Takács, Imre (2005). 'A New NATO Member's Perspective: Hungary's Army and Homeland Security' *Connections* Volume 4, Issue 3.
Varga, Gergely (2011). 'Security Sector Reform: Hungarian Experiences in the Defense Sector' *International Issues & Slovak Foreign Policy Affairs* Volume 20, Issue 3.
Vass, Mária (2002). 'Projecting Stability: Hungary's Role in Central and Southeastern Europe' *International Spectator* Volume 37, Issue 1.
Végh, Zsuzsanna (2015). 'Hungary's "Eastern Opening" Policy toward Russia' *International Issues & Slovak Foreign Policy Affairs* Volume 24, Issue 1–2.
Yaniszewski, Mark (2002). 'Post-Communist Civil-Military Reform in Poland and Hungary: Progress and Problems' *Armed Forces & Society* Volume 28, Issue 3.
Zord, Gábor (2007). 'New EW Challenges for Hungary' *Journal of Electronic Defense* Volume 30, Issue 3.

9 The strategy of Latvia[1]

The primary sources analysed in this chapter consist of the report on state defence policy of 2004, the national security concepts of 2005, 2008, 2011, 2016 and 2020, the defence concepts of 2012, 2016 and 2018, as well as the cyber defence concept of 2013.[2]

9.1 Historical background

Following the crusade against pagans in Northern Europe, Terra Mariana was established by the Christian orders. This state consisted of the present Latvia as well as the southern parts of Estonia. After centuries of rule by German, Poles, Russians and Swedes, Latvia declared independence from Russia in November 1918. However, the Latvians had to fight a war of independence against the Soviets before independence was recognized with the Treaty of Riga in August 1920. In May 1934, the initially introduced democracy was, after a *coup d'état*, replaced by an authoritarian regime. For slightly more than 50 years, from August 1940 to August 1991, Latvia shared the same experience of annexation, occupation and re-annexation as Estonia and Lithuania, that is by the USSR, by Nazi-Germany and then by the Soviets again (see, for example, Dreifelds 1996; Liddell-Hart 1997; Kasekamp 2010). According to the Latvian government, Latvian

> national security, economic welfare and development opportunities are determined by its geographical position, historic relations with neighbouring countries and nations, status of national economy, culture traditions and defence capabilities.
>
> (Latvian MoD 2004:7)

9.2 Strategic environment

According to the Latvian government, historic experiences as well as the current international security environment form the conditions for Latvian security. In 2004, the international security context was said to be characterised by increased expansion of asymmetrical threat such as international terrorism and ethnic

DOI: 10.4324/9781003298052-11

conflicts (Latvian MoD 2004). The Latvian Parliament, the Saeima, concluded that Latvia's membership in NATO and the EU had significantly

> changed the external security environment of Latvia by maximally reducing external military and political threats. There are opportunities open for Latvia, it had never experienced before, to take advantage of economical and political opportunities offered by the EU, and to become [an] equal co-operation partner of the other European countries within a relatively short period of time, thus enhancing [the] security of the whole European environment.
> (Saeima 2005:1)

The Latvian government concluded that there was no direct military threat either to Latvia or to any other Baltic State. The threats and challenges were considered as being mainly related to social stratification and the number of non-citizens in the country, economic security and the dependency on instable external markets and on nuclear power plants located in neighbouring countries (Saeima 2005). 'The internal security environment of Latvia is characterised by stability, the permanence of which is dependent on the potential development of known threats', the government concluded in 2008. Terrorism, including cyberterrorism, the spread of WMD and the tendency towards an increased arms race, were among the issues elaborated on by the government (Saeima 2008:16). In 2011, the Latvian government included threats caused by foreign intelligence and security services, threats to the unity of civil society, unfriendly countries in possession of nuclear weapons, the military conflict in Georgia and the military exercises conducted by Russia and Byelorussia in the elaborations. 'Currently Latvia has no direct military threats', the Parliament concluded (Saeima 2011:3). Notably, the Latvian government put emphasis on the domestic security environment, including the status of the Latvian language. 'Society in Latvia is split also by the attitude towards specific historical issues. The different attitudes influence cultural and political identity and global vision of specific groups of the society, which may lead to mutual disagreements', the Parliament argued (Saeima 2011:6). One year later, the cabinet took a rather global approach:

> The international security environment is changeable, complex and difficult to predict. Due to the globalisation process, Latvia has established closer ties with countries in different regions of the world. Consequently, any political, social, military or economic instability, even in the remote parts of the world can pose a direct or indirect threat to Latvia's national security.
> (Latvian MoD 2012:4)

Information technology 'threats or actions directed against the interests of national security in electronic information space are among the most topical national security threat factors', the government added (Latvian MoD 2013).

In 2016, the Parliament concluded that, since its approval of the previous national security concept in 2011, the international security environment had undergone significant changes. According to the Parliament, the new environment was 'characterised by instability and unpredictability, as well as formation of new conflicts, including those that are comparatively near the Latvian border'. The aggression in Ukraine 'fuelled by the Russian Federation' was described as presenting 'significant challenges to the security in Europe and global international order'. The Russian aggression had acted, the Parliament concluded, to 'significantly worsen the security within the Euro-Atlantic area and [has] created long-term effects on the national security of the Republic of Latvia' (Saeima 2016:3). The government also notified that Russia was undertaking military activities outside its territory in other contexts than Ukraine, for example in Syria. 'Russia is prepared to reach its goals regarding its neighbouring countries by any means, including the use of military force to enforce its foreign policy and security orientation', the government concluded. In addition, the intensified threat of terrorism and migration in the NATO and EU region, as well as the consequences of the economic crisis, was observed with specific concerns (Latvian MoD 2016:4). In 2020, the government observed that insufficient

> military spending by European countries continues to affect the ability to respond adequately to both European military threats and crises outside their territories. Although the US political will to maintain world order is waning, European countries' military dependence on the US remains in both collective defense and international operations.
>
> (Saeima 2020:3)

9.3 Ends

An initial strategic goal of the Latvian government after the end of the Cold War was to become a full member of NATO as well as of the EU. 'After becoming a full-fledged member of international organisations, the Latvian state and population can be confident about their future and contribute to the development of the country with determination', the government explained. Consequently, three core objectives were announced upon becoming a member of these organisations: 'provide the Latvian defence, fulfil the obligations of the NATO and the EU states, and participate in the international military operations'. Defending the Latvian national sovereignty and territorial integrity as well as the Latvian population against military aggression was the overarching end of the defence policy. In addition, the government presented its goal 'to facilitate the stability and security within Europe, strengthen the EU crisis management capabilities and contribute to the transatlantic co-operation' (Latvian MoD 2004:8–9 and 11).

Defending national independence, the democratic system, territorial integrity, human rights, domestic security and the welfare of people, the free development of the society, and the political and economic stability were all expressed as fundamental ends by the Latvian Parliament. Preventing threats and potential risk factors was also declared as a core objective. The national security interests were also said to include the preconditions required for long-term development of the

Latvian state and society. These preconditions included 'preserved language and cultural identity of the residents of Latvia, maintained defence system, developed scientific and technical potential, sustainable environmental development, ensured and developed critical infrastructure, telecommunications and security of information technology'. Good neighbouring relations and the increased development of democratic processes beyond the Baltic Sea region, especially 'considering the region of the CIS states' were also expressed as national interests (Saeima 2005:2).

The independence of the state, its constitutional system and territorial integrity, as well as the free development, welfare and stability of the society, the preservation of the language and culture of the population of Latvia, and the strengthening of solidarity among allied countries, were all considered key objectives of the Latvian government. In addition, the promotion of further cooperation in the Baltic Sea region, the strengthening of democratic statehood in neighbouring countries, as well as enhancing and increasing the functionality of organisations in which Latvia was a member, were mentioned as strategic ends of Latvia (Saeima 2008). In 2011, the Parliament repeated the independence of the country, its constitutional system and territorial integrity as overarching ends. In addition, the 'perspective of development for public freedom, welfare and stability' was mentioned in this context (Saeima 2011:2). In 2012, the cabinet stated that the key tasks of the national defence were to

> prevent, defeat and overcome potential national threats, to guarantee statehood, the capacity and continuation of state power and existing order, to contribute to international operations, to support civil society and engage in other emergency tasks in accordance with national laws and international agreements.
>
> (Latvian MoD 2012:5)

In 2020, the Parliament decided that the aim of national defence was 'to prevent and overcome possible threats to the state, to guarantee statehood, the capacity and continuity of state power and equipment, as well as to promote a responsible attitude of all Latvians towards the state and its security' (Saeima 2020:5).

9.4 Means

Soon after becoming a member of both NATO and the EU, the Latvian government announced its priorities when developing the country's armed forces. Modern command and control systems, efficient military combat capabilities including weapons for anti-aircraft and anti-tank defence, sophisticatedly equipped and armed units, the capacity to counter NBC weapons, readiness to participate in international military operations, establishing modern training bases, as well as meeting the requirements for receiving the support from the allies were at the top of the priority list. 'Latvia continues to improve its host-nation's support system', the government declared as it presented its decision of a 'gradual transition to

108 *The empirical exploration*

fully professional armed forces by 2006'. The government made clear that Latvia was dependent on the assistance of other NATO members to ensure full control and defence of Latvian air space. Regarding the ground forces, focus was on developing a brigade for national defence and on ensuring permanent participation of one battalion in NATO's operations (Latvian MoD 2004:10–11).

The bulk of the army was the brigade with its three infantry battalions. In addition, a special task unit was organised including rangers and combat divers. The National Guard was organised in two regions, each including eight infantry battalions and two specialised battalions for air defence and host nation support, respectively. The naval forces were organised in the War Ship Flotilla and the Coast Guard Ship Flotilla. The focus of these forces was on mine clearance, surveillance, as well as search and rescue operations. The Latvian air force was organised in the aviation squadron, the air defence wing and the air surveillance squadron. The main equipment consisted of transport aircraft, the RBS-70 mobile air defence weapon system, as well as long-range and medium-range air surveillance radars (Latvian MoD 2004). The Latvian Parliament stressed the necessity developing means for the defence of the Latvian territory as well as for contributing to international peacekeeping missions abroad. 'Transition from conscript service to professional military service, as well as the National Guard Reform is the main steps for achieving' these capabilities, the Parliament concluded (Saeima 2005:3). In 2011, the Parliament announced that adequate resources

> shall be dedicated for performing defence tasks and strengthening of [Latvian armed forces] battling capabilities. At the same time, developing abilities of [the armed forces], Latvia will attempt to participate in advantageous multinational projects, as well as to contribute to the economy of Latvia, involving Latvia's entrepreneurs in tenders organized by the Ministry of Defence, attracting NATO investments and supporting development of new technologies required for defence.
>
> (Saeima 2011:3–4)

The organisation of joint military training with other NATO member states on Latvian soil was considered as being of crucial importance. Consequently, the HNS system was given priority when developing the armed forces (Saeima 2011). Notably, the cabinet presented slightly different priorities. The development of the land force, 'which forms the core of the [armed forces'] capabilities, is a key priority for the [armed forces'] capability development and distribution of resources', the MoD argued. In order to increase the effectiveness of the armed forces, the cabinet also stressed the importance of ensuring a long-term national defence funding mechanism. This 'would provide for a gradual increase of national defence funding each year, so that it would reach 2% of the gross domestic product in the future. It would be necessary to achieve this objective by 2020', the cabinet declared (Latvian MoD 2012:3). Professionally prepared and trained, expeditionary and multifunctional armed forces were considered a precondition. 'Priority is the quality of capabilities, not the size of forces', the cabinet declared

Table 9.1 Main military resources of Latvia.

	2000	2010	2020
Army brigade	1	1	1
MBT	3	3	3
ACV/APC/IFV	13	0	0
Combat aircraft	0	0	0
Principle surface combatant	9 patrol craft	5 patrol craft	5 patrol craft

Source: International Institute for Strategic Studies (2000, 2010, 2020).

(Latvian MoD 2012:6). Hence, the number of professional soldiers was not to exceed more than 5,500, while the personnel strength of the National Guards, civilian employees and the reserve soldiers of the armed forces altogether was not to exceed 17,000. Regarding the contributions to international military operations, the cabinet declared that the Latvian armed forces had to ensure the ability to deploy and permanently sustain

> one platoon-level unit in an area of operations 15 000 km from Latvia's borders, one company-level unit at a distance of 5000 km, two company-level units with organic combat support and combat service support capabilities at a distance of 3000 km.
>
> (Latvian MoD 2012:8)

Contributions to both NATO and the EU high readiness forces were also considered important. However, only participation in the former was given an exact number, that is 390 soldiers. In addition, the Latvian ground forces were to contribute NATO-led operations with 'one infantry battalion for up to six months without rotation or with subunits not exceeding two infantry companies or equivalent size specialized subunits for up to six months with rotation'. The naval forces were to be able to contribute to mine countermeasures operations, but no exact number of vessels was presented. The government also announced plans for regular participation in EUBG, 'taking into account the involvement in other international operations' (Latvian MoD 2012:8). In 2016, special attention was paid to ensuring efficient intelligence, air defence and anti-tank capabilities, as well as tactical mobility (Latvian MoD 2016). Notably, in 2020 the two National Guard regions were referred to as brigades (Saeima 2020).

9.5 Ways

'Taking into account the lessons learned from history, the Latvian initial security and defence policy [after regaining independence] was based on the concept of the territorial defence', the government explained. The government stressed the necessity of a multilateral approach and claimed that the 'principle of the collective defence creates a prerequisite for Latvia to concentrate its efforts and resources to [. . .] ensure the Latvian participation in the international operations,

110 *The empirical exploration*

as well as increase its defence capabilities'. The government announced its ambition to expand the international cooperation with not only NATO and the EU but with the other Baltic States and the Baltic Sea region countries as well. At the same time, the government made clear that it considered the membership to NATO 'to be the most efficient way to secure [Latvia's] sovereignty and security' (Latvian MoD 2004:8–9 and 12). In 2004, Latvian armed forces had an infantry company of about 100 troops deployed to KFOR and another company deployed to the US-led Operation Iraqi Freedom in Iraq. In addition, the government pledged its willingness to contribute to the EU-led operation in Macedonia. 'Participation in international organisations is the means to strengthen the Latvian independence, security and economic growth, as well as to serve the state's national interests in the world', the Latvian Parliament declared (Saeima 2005:2). 'Latvia's participation in international peacekeeping operations has given the state an opportunity to become more widely involved in global political processes and to receive allied support in implementing national security interests', the Latvian Parliament concluded (Saeima 2008:1).

The Parliament stressed the importance of Latvia's participation in EU-led international missions such as the EU Monitoring Mission in Georgia (EUMM Georgia) and the EU Police Mission in Afghanistan (EUPOL Afghanistan). Participation in NRF was also considered important. Contributions to all these missions have 'provided opportunity for the country to [be involved] in global political processes and receive allies' support for implementation of national security interests', the Parliament argued (Saeima 2011:2). By strengthening NATO's collective defence and by contributing to international security ' both bilaterally and multilaterally, Latvia reduces the risk of external military aggression or an outbreak of other national threats, and, if necessary, is ready to ensure effective deterrence measures', the government argued (Latvian MoD 2012:3). Following Russia's military assault on Ukraine, the Latvian government put more emphasis on border control. 'Latvia can defend itself even when facing an adversary with military superiority. Everything is determined by the nation's willpower and the military strategy employed', the government declared. Obviously, this strategy included a substantial allied military presence in Latvia. 'Deterrence benefits from the continuous rotational presence of the allied forces', the government argued (Latvian MoD 2016:10–11). The US was declared being the major strategic cooperation partner of Latvia, and the American military presence in Latvia was especially welcomed. 'In order to achieve more efficient involvement of the US in Latvian security and defence activities, Latvia must cooperate in the military industrial sector by means of defence procurements; demonstrate solidarity by participating in US led military operations; develop military interoperability with US', the government concluded (Latvian MoD 2016:15).

In 2018, the government stressed the importance of individual willingness to resist occupation. 'Every inhabitant of Latvia should be given a specific role and position in the resistance', the government argued. If Latvia 'would fully or partly lose legitimate control over its territory during a crisis or armed conflict, resistance would be one of the ways to protect Latvia', the government announced (Latvian MoD 2018:2). In the event of massive military aggression, the Latvian armed forces were intended to weaken and inflict maximum damage on the opponent. In case of occupation of Latvian territory, the armed forces organise 'a public resistance movement' (Saeima 2020:6).

9.6 Conclusions: Latvian strategy

We find it reasonable to conclude that the Latvian alignment strategy corresponds to the criteria of a *multiple-courting* hedging strategy. Militarily, both NATO and the EU have over the past two decades been the focus. So have other organisations such as the OSCE and the UN as well. Only during the Trump administration has the US paid specific attention. Notably, NATO has been the main focus regarding the military aspects of security. However, the Latvian government has continuously connected the economic aspects and the financial conditions to its elaborations. Consequently, the EU has been regarded as a crucial instrument not only regarding the other aspects of security but also as a precondition for Latvia's economic developments and hence indirectly for its ability to allocate resources to the armed forces. Arguably, the ends have mainly been about the *survival* of the Latvian state. Notably, this has been the case also before the increased military threat from Russia. Not least has the Russian minority, which constitute about a quarter of the citizens, been of great concern in this regard.

When it comes to the means, we conclude that the Latvian government has had a clear focus on *national defence*. Despite the expressed ambitions regarding expeditionary capacity, providing platoons or occasionally a company for international operations is far from impressive. Since two out of three brigades clearly have tasks related to territorial defence, since the air force lacks combat aircraft and hence is focusing on air surveillance rather than flying, and since all services have HNS as a key task, we find our conclusion reasonable. These facts also provide arguments for the *multilateral approach* regarding the ways. Simply put, Latvia has no other options. With a population of fewer than 2 million, of which fewer than two-thirds are ethnic Latvians, the dependency of allies is obvious.

Table 9.2 Latvian strategy.

Alignment strategy	Military strategy		
	Ends	Means	Ways
Multiple-courting	Survival	National defence	Multilateral approach

We agree with Airis Rikveilis when arguing that the Latvian strategic elite considers the cooperation among the three Baltic States being of crucial importance. We also find his observations regarding developing a professional all-volunteer service in line with our conclusions. However, we do not agree with his claim that participation in international missions has been a cornerstone of the Latvian defence policy. His arguments regarding gaining military skills and international prestige simply do not correspond with the level of Latvian contributions. Rather, we find participation a necessity receiving allied attention (Rikveilis 2007; see also Kasekanp 2020). Consequently, we agree with Kārlis Kresliņš and his colleagues when arguing that Latvia has no other options in matters of defence, policy and strategy (Kresliņš *et al.* 2011).

Notably, Toms Rostoks and Nora Vanaga also argue that Latvia's key priority in the defence sector before the Ukraine crisis has been to contribute to international operations in Iraq and Afghanistan. Obviously, we do not agree. They may be right when claiming that these contributions have been seen as sufficient to ensure allied support for Latvia in the event of a military conflict. As previously stated, we do not think that Latvian participation has been impressive enough in this regard. However, we do agree that Latvia has negligible self-defence capabilities and that the concerns about the large Russian-speaking minority in Latvia have a great impact on the elaborations (Rostoks and Vanaga 2016; see also Paljak 2013). Neither do we agree with Māris Andžāns and Viljar Veebel when claiming that, after joining NATO, Latvia has neglected its territorial defence and focused on expeditionary capabilities instead. Our findings indicate the opposite (Andžāns and Veebel 2017). We rather support the findings of Valdis Otzulis and Žaneta Ozoliņa regarding the emphasis put on host nation support (Otzulis and Ozoliņa 2017).

We agree with Raimonds Rublovskis when arguing that the military cooperation among the Baltic States mainly has taken place at the tactical level and that all three countries are dependent on and the arrangements within NATO when it comes to the operational and strategic levels. In addition, we find his observation that the Nordic-Baltic military cooperation cannot be a substitute for NATO's security guarantees and the military involvement of the US (Rublovskis 2014). We do not agree with Dovile Jakniunaite when arguing that the situation in Ukraine and the response of NATO and the EU regarding the Russian aggression has constituted a 'matter of degree' in the Latvian security policy (Jakniunaite 2016). We rather agree with Andžāns and Veebel when claiming that, because of the Ukrainian crisis, NATO allies have more military solidarity with Latvia (Andžāns and Veebel 2017; see also Takacs 2017). We also agree with Živilė Marija Vaicekauskaitė when arguing that, after 2014, Latvia has strengthened its national defence capabilities, increased its defence spending and paid more attention to bilateral defence cooperation with the US (Vaicekauskaitė 2017; see also Vanaga 2019; Veebel 2019). Consequently, we agree with Erik Männik when concluding that 'Latvia valued very highly its relationship with the US and prioritised cooperation with the sole remaining superpower' (Männik 2013:24).

Notes

1 We would like to express our gratitude to Lieutenant Colonel Mikael Nordmark, Embassy of Sweden, Riga, for his support.
2 Andris Šķēle of the People's Party (TP) served as prime minister 16 July 1999–5 May 2000. Andris Bērziņš of the Latvian Way (LC) served as prime minister 5 May 2000–7 November 2002. Einars Repše of the New Era Party (JL) served as prime minister 7

November 2002–9 March 2004. Indulis Emsis of the Green Party (LZP) served as prime minister 9 March–2 December 2004. Aigars Kalvītis of the TP served as prime minister 2 December 2004–20 December 2007. Ivars Godmanis of the LC served as prime minister 20 December 2007–12 March 2009. Valdis Dombrovskis of the JL, to become the Unity (V), served as prime minister 12 March 2009–22 January 2014. Laimdota Straujuma of the V served as prime minister 22 January 2014–11 February 2016. Māris Kučinskis of the Liepāja Party (LP) served as prime minister 11 February 2016–23 January 2019. As of 1 January 2022, Arturs Krišjānis Kariņš of the New Unity (JV) had served as prime minister since 23 January 2019. See www.mk.gov.lv/en.

Bibliography

Andžāns, Māris and Viljar Veebel (2017). 'Deterrence Dilemma in Latvia and Estonia: Finding the Balance between External Military Solidarity and Territorial Defence' *Journal on Baltic Security* Volume 3, Issue 2.

Dreifelds, Juris (1996). *Latvia in Transition*. Cambridge: Cambridge University Press.

International Institute for Strategic Studies (IISS) (2000). *The Military Balance 2000–2001*. Oxford: Oxford University Press.

——— (2010). *The Military Balance 2010*. London: Routledge.

——— (2020). *The Military Balance 2020*. London: Routledge.

Jakniunaite, Dovile (2016). 'Changes in Security Policy and Perceptions of the Baltic States 2014–2016' *Journal on Baltic Security* Volume 2, Issue 2.

Kasekamp, Andres (2010). *A History of the Baltic States*. Basingstoke: Palgrave Macmillan.

——— (2020). 'An Uncertain Journey to the Promised Land: The Baltic States' Road to NATO Membership' *Journal of Strategic Studies* Volume 43, Issue 6–7.

Kresliņš, Kārlis, Aleksandrs Pavlovičs and Inese Kresliņa (2011). 'Defence of the Latvia: Past, Present and Future' *Baltic Security & Defence Review* Volume 13, Issue 2.

Latvian Ministry of Defence (MoD) (2004). *Report on State Defence Policy and Armed Forces Development*.

——— (2012). *State Defence Concept*.

——— (2013). *Cyber Defence Unit Concept*.

——— (2016). *National Defense Concept*.

——— (2018). *Comprehensive National Defense*.

Latvian Parliament (Saeima) (2005). *National Security Concept*.

——— (2008). *National Security Concept*.

——— (2011). *National Security Concept*.

——— (2016). *National Security Concept*.

——— (2020). *National Security Concept*.

Liddell-Hart, Basil (1997). *History of the Second World War*. London: Papermac.

Männik, Erik (2013). 'The Evolution of Baltic Security and Defence Strategies' in Tony Lawrence and Tomas Jermalavičius (eds). *Apprenticeship, Partnership, Membership: Twenty Years of Defence Development in the Baltic States*. Tallinn: International Centre for Defence Studies.

Otzulis, Valdis and Žaneta Ozoliņa (2017). 'Shaping Baltic States Defence Strategy: Host Nation Support' *Lithuanian Annual Strategic Review* Volume 15, Issue 1.

Paljak, Piret (2013). 'Participation in International Military Operations' in Tony Lawrence and Tomas Jermalavičius (eds). *Apprenticeship, Partnership, Membership: Twenty Years of Defence Development in the Baltic States*. Tallinn: International Centre for Defence Studies.

Rikveilis, Airis (2007). 'Strategic Culture in Latvia: Seeking, Defining and Developing' *Baltic Security & Defence Review* Volume 9, Issue January.

Rostoks, Toms and Nora Vanaga (2016). 'Latvia's Security and Defence Post-2014' *Journal on Baltic Security* Volume 2, Issue 2.

Rublovskis, Raimonds (2014). 'Latvian Security and Defense Policy within the Twenty-First Century Security Environment' *Lithuanian Annual Strategic Review* Volume 12, Issue 1.

Takacs, David (2017). 'Ukraine's Deterrence Failure: Lessons for the Baltic State' *Journal on Baltic Security* Volume 2, Issue 1.

Vaicekauskaitė, Živilė Marija (2017). 'Security Strategies of Small States in a Changing World' *Journal on Baltic Security* Volume 3, Issue 2.

Vanaga, Nora (2019). 'Latvia's Defence Strategy: Challenges in Providing a Credible Deterrence Posture' in Nora Vanaga and Toms Rostoks (eds). *Deterring Russia in Europe: Defence Strategies for Neighbouring States*. Abingdon: Routledge.

Veebel, Viljar (2019). 'Researching Baltic Security Challenges after the Annexation of Crimea' *Journal on Baltic Security* Volume 5, Issue 1.

10 The strategy of Lithuania[1]

The primary sources analysed in this chapter consist of the DWPs of 2002, 2006 and 2017, the NSSs of 2002, 2012 and 2017, the strategic guidelines of 2008, 2012, 2014 and 2015, the military strategies of 2012 and 2016, as well as the threat assessments of 2014, 2017 and 2020. In addition, reports on activities, forecasts of future global security developments and documents on cyber security have been studied.[2]

10.1 Historical background

Crusading military orders, such as the Teutonic Knights, were established during the early thirteenth century in Riga and in Prussia. In the mid-thirteenth century, Pope Innocent IV issued a papal bull proclaiming the creation of the Kingdom of Lithuania. This status was soon lost, but the country remained unified as a Grand Duchy. Lithuania was engaged in constant warfare with the Christian orders, Poland and the Kiev-Russians. The Polish-Lithuanian Union was a relationship between the Kingdom of Poland and the Grand Duchy of Lithuania that lasted from 1385 to 1569 when the Polish-Lithuanian Commonwealth was established. The Commonwealth existed until the late eighteenth century when it was portioned by the Habsburg Monarchy, Prussia and Russia. During its four centuries of existence, the Commonwealth fought several wars. Austria, Brandenburg, Bohemia, Denmark, Hungary, the Ottomans, Russia and Sweden were hence the main adversaries. In February 1918, Lithuania declared independence from Russia. As in the other Baltic States, the Lithuanians had to fight a war of independence against the Soviets before the sovereignty was recognized with the Treaty of Moscow in July 1920. In order to gain independence, Lithuania also had to fight a war against Poland. This war ended with Poland's annexation of the Vilnius region in 1922. In December 1926, the initially introduced democracy was, after a *coup d'état*, replaced by an authoritarian regime. From August 1940 to August 1991, Lithuania shared the same experience of Soviet, Nazi-German and renewed Soviet annexation and occupation as Estonia and Latvia (see, for example, Liddell-Hart 1997; Kasekamp 2010; Eidintas *et al.* 2013).

DOI: 10.4324/9781003298052-12

10.2 Strategic environment

At present, Lithuania 'does not observe any immediate military threat to national security and as a result does not regard any state as its enemy', the Parliament argued in the NSS of 2002. The Parliament nevertheless observed that non-military challenges, dangers and threats

> arise as a consequence of globalisation, therefore, individual states cannot respond to them alone. Such transnational factors as terrorism, organised crime, arms proliferation, drug traffic, the illegal migration, and the spread of epidemics defy state borders and become international security challenges, dangers and threats. The probability that they will continue to spread is increasing.
>
> (Seimas 2002:2–3)

Political extremism, social terrorism and activities of foreign intelligence agencies within Lithuania, as well as the proliferation of WMD, environment pollution, and uncontrolled migration, were also mentioned as potential risks. 'While the likelihood of direct military confrontation in the region is low, demonstrations of military force, provocations, and the threat to use force remain a danger to the security', the Parliament argued. The Parliament admitted that it was 'particularly interested in political, social, economic and ecological stability in the Kaliningrad region' (Seimas 2002:4 and 10). The government argued in similar ways:

> Although there is little likelihood of military confrontation in the region, this possibility cannot be rejected altogether. Besides, the demonstration of military power, initiating provocations and mock internal conflicts as well as implementation of other means of exercising economic and political pressure or threatening to use force continue to endanger Lithuania's security.
>
> (Lithuanian MoD 2002:7)

The overwhelming dependence of Lithuania on the strategic resources and energy supplies of one country, as well as the increasing differences in living conditions between different social groups within Lithuania, were also considered as risks (Lithuanian MoD 2002). Four years later, the government observed that although new unconventional security challenges were replacing 'traditional threats of armed aggression against the territory of sovereign states, the use of military force is not declining. Terrorist attacks in different countries and regions have reached such a level that the only way to prevent them is to use military force', the government concluded. The nature of the conflict itself was considered changing. 'Contemporary conflicts are more asymmetrical, open combat is more intense although shorter in duration, and battlefields are shifting from an open space into densely populated cities', the government observed. Moreover, it was considered highly probable that 'in the future, due to a growing economic differentiation between countries, social unrest or resistance to democratisation processes by authoritarian regimes, civil wars and ethnic clashes may impact entire regions' (Lithuanian MoD 2006:7). When presenting the new NSS in 2012, the Parliament elaborated on some external risks, dangers and threats. Nuclear energy facilities in Lithuania's neighbourhood disregarding international nuclear energy safety standards was such a topic. The activities of other states against Lithuania in order to have impact on the political system, military capabilities, social and economic life and/or cultural identity were another, and information and/or cyber attacks a third (Seimas 2012). 'Although contradictory tendencies prevail in the international security environment, the Lithuanian security environment remains rather favourable and predictable', the government nevertheless concluded (Lithuanian MoD 2012a:3). In 2013, in its first attempt ever to systematically assess the future security environment, the government predicted that, over the upcoming two decades

> multipolarity should gradually establish itself in the international system, reflecting the shift of power towards the East. Apart from the US, which will maintain its leading positions, the EU, fostering the ambitions of a single actor in the international system, and Russia, making every endeavour not to lose the status of one of the major world powers, China and India will definitely pave their way to the list of the most powerful states in the world.
> (Lithuanian MoD 2013:46)

In 2014, the government concluded that the

> military occupation and annexation of Ukraine's Crimea Peninsula along with actions intended to disintegrate Ukraine demonstrated that Russia has no respect for independence and territorial integrity of its neighbouring countries, and is inclined to violate international law brutally.
> (Lithuanian MoD 2014a:3)

The government argued that the likelihood of a conventional armed aggression against Lithuania and/or other NATO members in the Baltic Sea region no longer could be regarded only as theoretical elaborations. The conclusion was based on Russia's growing military power (Lithuanian MoD 2016). 'Threats and challenges posed to Lithuanian national security by Russian and Belarusian intelligence and security services will be growing in intensity', the government predicted (Lithuanian State Security Department (SSD) 2017:39). The government also observed 'that cyberspace has been increasingly used as a separate military space or as a tool of hybrid warfare' (Lithuanian MoD 2018:8; see also Seimas 2018). Moreover, ongoing 'malicious use of Russian and Chinese cyber capabilities is being observed in Lithuanian cyberspace' (Lithuanian SSD 2020:6).

10.3 Ends

'Lithuania perceives its security as preserving its sovereignty and territorial integrity, internal security and order, democratic foundations, economic security of all

118 *The empirical exploration*

legal entities and population and protection of its natural environment', the Parliament declared in the NSS issued in 2002. Joining and participating in international organisations was hence considered essential. 'In this regard NATO and the EU occupy a special place among such institutions with Lithuanian membership in each as the highest priority', the Parliament announced (Seimas 2002:1–2). The Parliament declared sovereignty, territorial integrity, democratic constitutional order, respect and protection of human and civil rights and freedoms, as well as peace and prosperity of the state, as Lithuania's vital interests. Primary interests were also listed: 'global and regional stability; freedom and democracy in Central and Eastern Europe and the Baltic States; open and predictable security policy of all countries in the Euro-Atlantic area; ensuring alternative energy supplies and supply of resources that are of strategic importance; and a region free of environmental dangers'. Fostering citizenship and patriotism was also mentioned as an objective (Seimas 2002:3). Additional primary interests were said to be friendly relations with neighbours and increased Euro-Atlantic integration. 'The basis of state national security is a strong, healthy and confident nation, confiding in its state', the government declared (Lithuanian MoD 2002:5). 'The main objective of Lithuania's defence policy remains unchanged', the government declared four years later (Lithuanian MoD 2006:15).

However, six years later, some additional ends were presented. Some of these ends were related to security, that is economic and energy security, information security, cyber security and infrastructure security. The Parliament also mentioned ethnic and cultural identity, as well having respect from the international community, as ends. Moreover, having 'credible national defence capabilities, supported by national defence financing which meets the needs of defence and commitments to allies' was also presented as an end (Seimas 2012:4). In 2017, the Parliament emphasised the need for strengthening the will of the population to defend the state as well as their total preparedness to resist an armed aggression. Strengthening patriotism, fostering the historical memory and protecting 'the Lithuanian language, [so as to] ensure preservation of the ethnic identity as well as tangible and intangible cultural heritage' of the Lithuanian nation were hence articulated as central objectives (Seimas 2017:14).

10.4 Means

In the white paper on defence presented in 2002, the government announced its ambitions transforming the Lithuanian armed forces. Hence, the land forces were given priority. The motorised infantry brigade was to be reorganised into a reaction brigade consisting of two mechanised battalions and one motorised battalion. Moreover, three independent battalions, one light infantry, one jaeger and one engineer, were to provide additional strength. Two territorial regions were also to be organised, the western with three battalions and the eastern with one. The National Defence Volunteer Forces were to provide additional units designed for territorial defence to the regions. The bulk of the air force was to consist of two airbases and one air defence battalion. Ten helicopters, 11 transport aircraft and six training aircraft were the flying resources. The navy was to sail two light frigates, two minesweepers and three patrol vessels (Lithuanian MoD 2002). 'Well prepared expeditionary forces are needed', the government concluded in 2006 and presented a list of potential contributions. A battalion task group, an infantry company, two engineer platoons, a special operations squad, a mine countermeasure ship, a transport aircraft and a logistic battalion were listed. The two regional commands were disbanded, and so was one of the airbases. One of the battalions previously detached to the regional commands was transferred to the brigade. In addition, a logistic command was set up while the special operations forces were organised in a new command, including the jaeger battalion (Lithuanian MoD 2006:11).

In 2008, the government had to take the difficult economic situation of Lithuania and the reduced financing of the armed forces into account and hence presented priorities. The first priority was to 'prepare for and participate in international operations, paying special attention to the NATO-led operation in Afghanistan', and the second was to 'prepare for and participate in NATO NRF and the EU Battlegroups stand-by' (Lithuanian MoD 2008:1). In 2012, the government announced an ambition having 50 percent of the land forces prepared to be deployed outside the territory of Lithuania. Notably, the ambition differed depending on context. To international operations led by NATO, up to one infantry battalion battle group, one special operations forces squadron, one mine countermeasures vessel (MCMV) and/or one light transport aircraft were to be committed. To operations conducted by the EU, one company-sized infantry unit, with combat support and combat service support elements, and/or a combined specialized unit were to be contributed. To operations led by the UN or the OSCE, platoon-sized units as well as individual officers were to be provided. Hence, developing the land forces was a priority. Having a fully manned brigade-sized unit, supported by combat support and combat service support units, capable of sustaining itself independently was the aim of the continued developments. Provision of the HNS to NATO missions, as well as preparation for and participation in NRF and EUBG, were also given priority (Lithuanian MoD 2012b).

In 2014, the government prioritised three areas of bilateral cooperation when developing the capabilities of the armed forces. First, military cooperation with the US strived to enhance the US military presence in the Baltic Sea region by organizing regular joint exercises. Second, in order to develop the already existing defence relations with Poland, the government intended to promote the establishment of a Lithuanian, Polish and Ukrainian brigade. Third, the affiliation of the Lithuanian brigade to the Danish division was to be developed (Lithuanian MoD 2014b). One year later, the government announced a reformation of the Lithuanian land force. The mechanised infantry brigade was hence to be reorganised, enabling the establishment of a new motorised infantry brigade. The 'top priority will be the mechanized infantry brigade "Iron Wolf", while the motorised infantry brigade will be second priority', the government declared (Lithuanian MoD 2015:3). In order to man the Lithuanian armed forces, the Parliament announced that a method using the mixed model of professional military servicemen, conscripts and national defence

120 *The empirical exploration*

Table 10.1 Main military resources of Lithuania.

	2000	2010	2020
Army brigade	2	1	2
MBT	0	0	0
ACV/APC/IFV	14	187	230
Combat aircraft	0	0	0
Principle surface combatant	2 corvettes, 4 patrol craft	1 corvette, 4 patrol craft	4 patrol craft

Source: International Institute for Strategic Studies (2000, 2010, 2020).

volunteers was to be implemented. This approach was also intended to 'enlarge and strengthen the prepared reserve of military personnel' (Seimas 2017:7). 'Lithuania must develop national defence capabilities which are sufficient for national defence, with the object ability to fight until Allied reinforcements are in place', the government argued. Consequently, a national rapid response force was established. Moreover, a new infantry and a new artillery battalion had been organised within the frames of the second brigade (Lithuanian MoD 2017:12).

10.5 Ways

Deterrence based on defence was declared to be the key approach in the NSS of 2002. Lithuania 'gives priority to conflict prevention, diplomacy, and international legal measures. Of particular importance is the priority given to participating in international crisis management', the Parliament declared. Lithuania 'considers NATO membership as a principal means of ensuring both internal and regional security and stability' the Parliament continued. The strategic partnership with the US was also given highest priority (Seimas 2002:7–8). 'As a future EU member state Lithuania supports the extension of the EU defence dimension, as far as it does not challenge the role of NATO', the government announced. 'Lithuania seeks accession to NATO, because only membership in NATO's collective defence system will assure long-term security and stability for Lithuania', the government explained (Lithuanian MoD 2002:6 and 8). The government nevertheless made clear that in the event of an armed attack,

> Lithuania shall not allow the enemy to attain its goals. [. . .] Resistance will not cease until the complete state sovereignty and territorial integrity is [sic] reinstated. [. . .] The potential aggressor must know that in attacking Lithuania it will encounter armed resistance as well as national opposition and shall incur huge losses disproportional with the desired objectives.
> (Lithuanian MoD 2002:8)

Surveillance of the airspace, air traffic control and providing HNS were the primary wartime tasks of the air force. The basic mission of the naval force was to control, protect and defend the territorial waters, including the exclusive economic zone, of Lithuania (Lithuanian MoD 2002). 'Lithuania does not consider it necessary for the EU to develop structures and capabilities that NATO already possesses', the government declared after entering both organisations. At the same time, the government concluded that active 'participation in operations enables Lithuania to assert its position in international organisations and have more influence in setting their agendas. Hence, the government mentioned not only NATO and the EU but also the UN and the OSCE, although the two former were given priority. Multilateralism was declared a core principle on which Lithuania's defence policy was based. The trilateral cooperation of the Baltic States was hence described as essential (Lithuanian MoD 2006:18–19). The 'defence of Lithuania is total and unconditional', the government declared. The defence of Lithuania 'is not subject to any conditions and no one is allowed to inhibit the right of the nation and every citizen to resist an aggressor, invader or anyone else threatening by force' the vital interests of the republic, the government continued. Ensuring a reliable deterrence together with the allies and, if deterrence fails, ensuring the defence of Lithuania individually as well as together with the allies were fundamental in the elaborations of the ways. In addition, contributing 'to collective defence actions outside the territory of Lithuania in case of aggression against any other NATO member' was also considered fundamental (Lithuanian MoD 2012a:6).

In 2016, the government stressed the necessity of preparing for both the individual and collective defence of Lithuania, even though 'NATO membership guarantees that in case of an attack, Lithuania will be defended by the armed forces of other NATO nations'. Promoting bilateral and multilateral military cooperation with especially the US, the Baltic and Nordic states, Poland, the UK, Germany and France was hence considered as a necessity rather than as an option (Lithuanian MoD 2016:6). 'In response to the changing nature of the threats, Lithuania has adopted a new defence concept', the government announced in 2017. The national rapid response force was to react to local armed incidents as well as border violations during peacetime (Lithuanian MoD 2017).

10.6 Conclusions: Lithuanian strategy

Arguably, the Lithuanian alignment strategy correspond to the *multiple-courting* hedging strategy. Over the past two decades, not only NATO but also the EU and the UN have been the focus militarily even if the former has been prioritised. In addition, the Lithuanian government has established several bi-, tri- and other multilateral arrangements. Hence especially the other Baltic States and Poland have received the greatest attention. Regarding the ends of the military strategy, the government presented a rather broad approach with different categories. Since most of the ends of the prioritised category, the vital interests, clearly have to do with *survival*, we conclude that this is at the core of the Lithuanian strategy. Despite the rather constant priority regarding the ends, we find a shift in the priority regarding the means. Initially, the government launched a transformation aimed at increasing the capacity for expeditionary warfare and contributions to international military operations. Hence, the concept of territorial defence was

122 *The empirical exploration*

abandoned, and units designed for national defence were dismantled. Notably, Lithuania participated in international operations with resources from all three services. Despite the economic difficulties, the international focus of the armed forces was prioritised.

Following the Russian war against Ukraine in 2014, the government shifted focus back to *national defence* and organised a second brigade. Over time, HNS and the ability to receive allied reinforcements have been the focus of the air force. Moreover, since the air force lacks combat aircraft and hence is focusing on air surveillance rather than flying, we find our conclusion reasonable. Clearly, a *multilateral approach* regarding the ways is a necessity, not an option. With a population of fewer than 3 million, of which about 85 percent are ethnic Lithuanians, the dependency on allies is obvious. The Lithuanian concept focused on deterrence, both nationally within the framework of total and unconditionally defence and collectively within the framework of NATO.

Table 10.2 Lithuanian strategy.

Alignment strategy	Military strategy		
	Ends	Means	Ways
Multiple-courting	Survival	National defence	Multilateral approach

We appreciate Andres Kasekamp (2020) describing and analysing Lithuania's and the other Baltic States' road to their NATO membership in 2004. Regarding Lithuania specifically, we agree with Renatas Norkus when observing that the country initially took a quite 'remarkable departure from the focus on national defence' (Norkus 2006:167; see also Männik 2013). However, contrary to Kasekamp, Männik and Norkus, we also explore the time after Russia's aggression against Ukraine. Consequently, we have been able to observe the refocus on national defence. Hence, the findings of Dovile Jakniunaite are in line with our own conclusions (Jakniunaite 2016; see also Vileikiene and Janušauskiene 2016). We also agree with Norkus that Lithuania tends to favour participating in campaigns led by NATO, the EU and the US. However, we do support Viktorija Cieminytė when arguing that the development of the Lithuanian armed forces had a focus on 'capabilities to effectively contribute to a full range of NATO's operations' (Cieminytė 2006:209; see also Paljak 2013). In this context, the findings of Eric de Bakker and Robert Beeres are interesting. They argue that Lithuania never fulfilled the commitment TO spend 2 percent of GDP on defence. Moreover, following the global financial crisis, Lithuania's expenditures on defence 'were cut quickly and radically' (de Bakker and Beeres 2012:20; see also Odehnal *et al.* 2020; Pavlíčková and Bartoszewicz 2020).

Valdis Otzulis and Žaneta Ozoliņa observed that Lithuania from 2014 gave priority to developing HNS and self-defence capabilities. Obviously, this is in line with our findings. However, Otzulis and Ozoliņa go one step further and argue that Lithuania's lack of necessary infrastructure 'may cause friction in receiving Allied forces' (Otzulis and Ozoliņa 2017:97). David Takács also noted the shift in focus from 2014, and his conclusions on the importance of deterrence measures supports our conclusions in this regard (Takacs 2017; see also Janeliunas 2019; Veebel 2019). Similarly, the findings of Živilė Marija Vaicekauskaitė on Lithuania's bi- and multilateral defence cooperation provide support to our arguments regarding the alignment strategy (Vaicekauskaitė 2017). Deividas Šlekys provide additional support in this regard. His findings on Lithuania's balancing act between national defence and participation in international military operations are interesting. Although we agree with his argument that the 'emphasis on territorial defence will tie up the majority of Lithuanian forces', we do not share his assumption that Lithuania will continue to be as active internationally as before 2014 (Šlekys 2017:54). Arguably, the findings of Ingrida Gečienė-Janulionė rather indicate that Lithuanian citizens prefer increased attention to the national defence (Gečienė-Janulionė 2018). According to David Scott, this attention may very well have to include China in the calculus of the traditional perceived Russian threat (Scott 2018).

Notes

1 We would like to express our gratitude to Commander Giedrius Valintelis, Embassy of Lithuania, Copenhagen, and Lieutenant Colonel Lars Eklind, Embassy of Sweden, Vilnius, for their support.
2 Andrius Kubilius of the Homeland Union–Lithuanian Christian Democrats (TS-LKD) served as prime minister November 1999–October 2000 and again November 2008–November 2012. Rolandas Paksas of the Liberal Union served as prime minister October 2000–June 2001. Algirdas Brazauskas of the Lithuanian Social Democratic Party (LSDP) served as prime minister July 2001–June 2006. Gediminas Kirkilas of the LSDP served as prime minister July 2006–November 2008. Algirdas Butkevičius of the LSDP served as prime minister December 2012–December 2016. Saulius Skvernelis served as an independent prime minister endorsed by Farmers and Greens Union (LVZS) December 2016–November 2020. As of 1 January 2022, Ingrida Šimonytė has served as an independent prime minister since December 2020. See https://lrv.lt/en.

Bibliography

Cieminytė, Viktorija (2006). 'Lithuania: Policy of Active Membership' *Baltic Security & Defence Review* Volume 8, Issue January.
de Bakker, Eric and Robert Beeres (2012). 'Comparative Financial Analysis of Military Expenditures in the Baltic States, 2000–2010' *Baltic Security and Defence Review* Volume 14, Issue 1.
Eidintas, Alfonsas, Alfredas Bumblauskas, Antanas Kulakauskas and Mindaugas Tamošaitis (2013). *The History of Lithuania*. Vilnius: Eugrimas.
Gečienė-Janulionė, Ingrida (2018). 'The Consequences of Perceived (In)Security and Possible Coping Strategies of Lithuanian People in the Context of External Military Threats' *Journal on Baltic Security* Volume 4, Issue 1.
International Institute for Strategic Studies (IISS) (2000). *The Military Balance 2000–2001*. Oxford: Oxford University Press.
——— (2010). *The Military Balance 2010*. London: Routledge.
——— (2020). *The Military Balance 2020*. London: Routledge.

Jakniunaite, Dovile (2016). 'Changes in Security Policy and Perceptions of the Baltic States 2014–2016' *Journal on Baltic Security* Volume 2, Issue 2.

Janeliunas, Tomas (2019). 'The Deterrence Strategy of Lithuania: In Search of the Right Combination' in Nora Vanaga and Toms Rostoks (eds). *Deterring Russia in Europe: Defence Strategies for Neighbouring States*. Abingdon: Routledge.

Kasekamp, Andres (2010). *A History of the Baltic States*. Basingstoke: Palgrave Macmillan.

——— (2020). 'An Uncertain Journey to the Promised Land: The Baltic States' Road to NATO Membership' *Journal of Strategic Studies* Volume 43, Issue 6–7.

Liddell-Hart, Basil (1997). *History of the Second World War*. London: Papermac.

Lithuanian Ministry of Defence (MoD) (2002). *White Paper on Lithuanian Defence Policy*.

——— (2006). *White Paper on Lithuanian Defence Policy*.

——— (2008). *Guidelines for 2009–2014*.

——— (2012a). *Military Strategy*.

——— (2012b) *Guidelines for 2012–2017*.

——— (2013). *The World in 2030*.

——— (2014a). *Assessment of Threats to National Security*.

——— (2014b) *Guidelines for 2014–2019*.

——— (2015). *Guidelines for 2016–2021*.

——— (2016). *Military Strategy*.

——— (2017). *White Paper on Lithuanian Defence Policy*.

——— (2018). *National Cyber Security Strategy*.

Lithuanian National Assembly (Seimas) (2002). *National Security Strategy*.

——— (2012). *National Security Strategy*.

——— (2017). *National Security Strategy*.

——— (2018). *Law on Cyber Security*.

Lithuanian State Security Department (SSD) (2017). *National Security Threat Assessment*.

——— (2020). *National Security Threat Assessment*.

Männik, Erik (2013). 'The Evolution of Baltic Security and Defence Strategies' in Tony Lawrence and Tomas Jermalavičius (eds). *Apprenticeship, Partnership, Membership: Twenty Years of Defence Development in the Baltic States*. Tallinn: International Centre for Defence Studies.

Norkus, Renatas (2006). 'Lithuania's Contribution to International Operations: Challenges for a Small Ally' *Baltic Security & Defence Review* Volume 8, Issue January.

Odehnal, Jakub, Jiří Neubauer, Lukáš Dyčka and Tereza Ambler (2020). 'Development of Military Spending Determinants in Baltic Countries: Empirical Analysis' *Economies* Volume 8, Issue 3.

Otzulis, Valdis and Žaneta Ozoliņa (2017). 'Shaping Baltic States Defence Strategy: Host Nation Support' *Lithuanian Annual Strategic Review* Volume 15, Issue 1.

Paljak, Piret (2013). 'Participation in International Military Operations' in Tony Lawrence and Tomas Jermalavičius (eds). *Apprenticeship, Partnership, Membership: Twenty Years of Defence Development in the Baltic States*. Tallinn: International Centre for Defence Studies.

Pavlíčková, Kristýna and Monika Gabriela Bartoszewicz (2020). 'To Free or Not to Free (Ride): A Comparative Analysis of the NATO Burden-Sharing in the Czech Republic and Lithuania' *Defense & Security Analysis* Volume 36, Issue 3.

Scott, David (2018). 'China and the Baltic States: Strategic Challenges and Security Dilemmas for Lithuania, Latvia and Estonia' *Journal on Baltic Security* Volume 4, Issue 1.

Šlekys, Deividas (2017). 'Lithuania's Balancing Act' *Journal on Baltic Security* Volume 3, Issue 2.

Takacs, David (2017). 'Ukraine's Deterrence Failure: Lessons for the Baltic State' *Journal on Baltic Security* Volume 2, Issue 1.

Vaicekauskaitė, Živilė Marija (2017). 'Security Strategies of Small States in a Changing World' *Journal on Baltic Security* Volume 3, Issue 2.

Veebel, Viljar (2019). 'Researching Baltic Security Challenges after the Annexation of Crimea' *Journal on Baltic Security* Volume 5, Issue 1.

Vileikiene, Eglė and Diana Janušauskiene (2016). 'Subjective Security in a Volatile Geopolitical Situation: Does Lithuanian Society Feel Safe?' *Journal on Baltic Security* Volume 2, Issue 2.

11 The strategy of Poland[1]

The primary sources analysed in this chapter consist of DWPs of 2001 and 2013, the NSSs of 2003, 2007, 2014 and 2020, the NDS of 2009, and the defence concepts of 2017. In addition, the NCSS of 2017, the foreign policy strategy of 2017 and the agreement on enhanced defence cooperation with the US of 2020 have been explored.[2]

11.1 Historical background

The Duchy of Poland was established about 960 and the Polish Kingdom in 1025. Three and a half centuries later, in 1385, the Polish Queen Jadwiga married Jogaila, the Grand Duke of Lithuania, hence uniting the two countries. During the four centuries before the unification, Poland fought several wars, mainly against the Holy Roman Empire (as well as against separate parts of the Empire), Hungary and/or the Kiev-Russians. In 1569, Poland and Lithuania established a commonwealth. During the following two centuries, the Swedish Empire became an additional main enemy. Between 1772 and 1795, this commonwealth suffered from three partitions between the Habsburg Monarchy, Prussia and Russia. After slightly more than a century, Poland re-emerged as a *de facto* independent state in November 1918 and as a *de jure* state following the Treaty of Versailles in June 1919. However, Poland had to fight a brief war with Czechoslovakia in 1919 and a war with the Soviets 1919–1921 before its sovereignty was ensured. Initially, Poland was a democracy, but in May 1926 Marshal Józef Piłsudski seized power and introduced an authoritarian government. Although Poland suffered annexation and occupation by Nazi-Germany and the USSR during WWII, troops loyal to the Polish government in exile continued to fight Germany throughout the war. Following a decision by the victorious Allies at the Potsdam Conference in August 1945, Poland's territory shifted westwards. The USSR annexed the former eastern parts of Poland, while former German territory east of the Oder–Neisse line was transferred to Poland as compensation. Contrary to the experiences of Hungary (1956) and Czechoslovakia (1968), the Polish civil resistance during 1980–1989, organised by Solidarity under the leadership of Lech Wałęsa, did not lead to a Soviet invasion (see, for example, Liddell-Hart 1997; Sanford 1999; Banaszkiewicz-Zygmunt and Olendzki 2000).

DOI: 10.4324/9781003298052-13

11.2 Strategic environment

Notably, in 2001, the Polish government took Poland's geographical position as a starting point in its elaborations on the strategic environment.

> In the event of war in Europe, the territory of Poland would play a vital role in securing strategic freedom for the European theatre of operations. The past periods of great military confrontations in this part of the world always brought tragic consequences for our country.
>
> (Polish MoD 2001:9)

However, in 2003, the Polish National Security Bureau (NSB) focused on new global challenges 'provoked by international terrorism and the proliferation of weapons of mass destruction, as well as the unpredictable policies of authoritarian regimes' and failed states. 'Economic backwardness, poverty, degradation of the natural environment, lethal disease epidemics, uncontrolled migrations and ethnic tensions tend to increasingly erode stability of the international system', the NSB concluded. Organized international crime and foreign special services were also mentioned as sources of increased concerns (Polish NSB 2003:2).

The memberships in NATO and the EU as well as the alliance with the US 'have ensured Poland a high level of security and have become one of the fundamental guarantees of its internal development and its international position', the government argued in 2007. At the same time, the government observed that the US's international position had weakened. Russia was accused of intensively attempting to reinforce its position. The worsening security situation in the Middle East, including Iran's nuclear ambitions and Islamic fundamentalism, was, the government concluded, exerting a negative influence on the security situation also in Europe. 'The use of energy resources as an instrument of political pressure by some states and the growing rivalry for energy carriers contribute to greater risks in this area', the government argued. Unresolved regional and local conflicts in general and specifically in Transnistria, South Caucasus and the Balkans were elaborated. Rouge states unable to control their territories, mass migrations, organised international crime and the proliferation of WMD were also considered as worrisome. 'The dependence of [the] Polish economy on supplies of energy resources [. . .] from one source is the greatest external threat to our security', the government declared. Regardless of all these challenges, the government concluded that in 'the foreseeable future the eruption of a large-scale armed conflict is unlikely' (Polish NSB 2007:2, 8 and 10). In 2009, the government concluded that the conflict in Georgia had 'demonstrated the topicality of traditional military threats and the importance of military force, also in Europe's backyard' (Polish MoD 2009:3–4). Four years later, the government concluded that military threats

> still remain significant. They may primarily have the nature of political and military crises provoked to exert a strategic pressure in the current politics, without crossing the brink of war. Such activities might take the form of rapid

development of military potential near the Polish borders, practical demonstration of strength or military blackmail. Direct armed threats, however, cannot be excluded.

(Polish MoD 2013:13)

In 2017, the government went even further, concluding that the scale of threats resulting from Russia had not been adequately assessed in the past:

> Russia is ready to destabilize the internal order of other states and to question their territorial integrity by openly violating international law. Russia's actions are often camouflaged and conducted below the threshold of an armed conflict. It is not unrealistic that Russia could incite a regional conflict and drag into it one or several NATO countries. Russia is also likely to provoke proxy wars in various parts of the world in order to exert pressure on the Western countries.
>
> (Polish MoD 2017:24)

The failure to respond to Russia's aggressive security policy revealed, according to the Polish government, the West's weakness. 'This ultimately led to war in Ukraine and undermined Europe's security architecture', the government concluded. Russia was accused being a revisionist power willing to resort to military means in order to achieve its goals (Polish MoFA 2017:2). The government warned that Russia

> is intensively developing its offensive military capabilities (including in the western strategic direction), extending Anti-Access/Area Denial systems *inter alia* in the Baltic Sea region [. . .] and conducting large-scale military exercises, based on scenarios assuming a conflict with the NATO member states, a rapid deployment of large military formations, and even the use of nuclear weapons.
>
> (Polish NSB 2020:6)

11.3 Ends

Defence against external crisis or threats of major conflict were, in 2001, declared to be the principal objectives of Polish Defence Policy. These objectives were said to encompass: (1) defending the Polish territory against armed aggression, (2) ensuring the inviolability of the Polish air and maritime space and land boundaries, (3) protecting all citizens of Poland, (4) participation in the common defence of the territories of other NATO members, and (5) supporting international organisations by responding to crisis situations with military stabilization measures (Polish MoD 2001). In 2003, the government declared the fundamental security objective to be safeguarding Poland's sovereignty and independence as well as border inviolability and territorial integrity. In addition, the security policy was said to promote the security of the citizens, human rights and fundamental freedoms, democratic order, stable conditions for Poland's civilisational and economic progress, well-being of the people, protection of national heritage and national identity, implementation of allied commitments, defence capability and interests of the Polish State.

(Polish NSB 2003:1)

The government described Poland as 'a sovereign and democratic state in Central Europe with significant population, political, military and economic potential' with three groups of national interests: (1) vital, (2) important, and (3) other significant. The first group involved guaranteeing the survival of the state and its citizens, preserving Poland's independence, sovereignty, and territorial integrity, and ensuring the security of Poland's citizens. The interests in the second group were guaranteeing that Poland 'develops civilizationally and economically, that conditions are created for the growth of a more prosperous society, for the development of science and technology and for a proper protection of its national heritage and identity'. The last group entailed ensuring that Poland 'maintains a strong international position and is capable of effectively promoting Polish interests abroad'. The significant interests also included 'strengthening the ability to operate and to be effective [in] the most important international institutions in which Poland participates'. The promotion of the image of Poland being a credible ally and building 'Poland's prestige in the international environment' were declared as strategic goals. Consolidating Poland's international position was also mentioned in this context (Polish NSB 2007:4–6 and 10). Ensuring 'that Poland occupies a strong international position and that Polish interests on the international arena are being effectively promoted' were mentioned as core objectives (Polish MoD 2009). 'The historical experience which has determined the Polish strategic culture [. . .] serve as a basis for the formulation of a catalogue of national interests and strategic objectives in the field of security', the government announced. (Polish MoD 2013:11). This catalogue included:

- possession of effective national security capacities ensuring readiness and ability to prevent threats, including deterrence, defence and protection against them, as well as elimination of their consequences;
- [a] strong international position of Poland and membership in reliable international security systems.

(Polish NSB 2014:10)

Strengthening the links between the Polish diaspora and the home country, and increasing the involvement of the diaspora in activities related to the promotion of Poland was another objective presented by the government (Polish NSB 2020).

11.4 Means

In 2001, the Polish government admitted the weakness of the Polish armed forces, which had to rely on outdated equipment originating from the USSR with low

130 *The empirical exploration*

combat value. Consequently, the government announced its ambition transforming the Polish military. The plans included reducing the size of the army to 150,000 soldiers within six years. 'One third will be trained and equipped to NATO standards. This will be a Polish Army worthy of its one thousand years tradition and at the same time an army of the XXI century, significantly modern, mobile and well equipped', Bronistaw Komorowski, the minister of national defence, declared (Polish MoD 2001:6). However, in 2003, the government stressed that the size, organisation and assets of the armed forces were to be continually adapted to not only the national defence requirements but also regarding allied and international commitments, as well as Poland's social-economic potential. 'As the nature of security threats evolves, static armed forces designed for territorial defence will be gradually phased out in favour of advanced, mobile, highly specialised units', the government announced (Polish NSB 2003:4).

In 2007, the government presented its ambition developing the armed forces for both national defence and expeditionary warfare. On the one hand, Poland was said to develop the combat readiness of its armed forces 'to ensure effective defence and protection of Polish borders within the framework of operations carried out independently or as part of collective defence, as well as outside its borders, pursuant to Article V of the Washington Treaty'. On the other hand, the Polish armed forces were to be ready to participate in 'operations of asymmetrical nature, including multinational, joint anti-terrorist operations'. Moreover, the armed forces were to 'participate in a significant way in NATO- or EU-led crisis response operations and to support such operations organized by the UN'. The transformation of the armed forces was hence primarily focused on the replacement of armaments and equipment, as well as on the reorganization of the force structures, in order to increase operational readiness and troop mobility (Polish NSB 2007:23). At the same time, the government announced that the

> size of the Armed Forces of the Republic of Poland will not be altered significantly in the nearest future. Reductions made during the past 20 years or so have brought the size of the armed forces to a level on which the continuation of this tendency may carry unwanted risk. However, [the] armed forces will gradually become professional.
>
> (Polish NSB 2007:24)

In 2009, the government stressed that the transformation of the armed forces was to ensure the capability 'to deter, protect and defend, and to execute tasks of collective defence' as well as 'to cooperate with Allied structures, while preserving the capability to effectively act alone' (Polish MoD 2009:12). The government announced an increased ambition regarding strategic transportation, striving for the capacity to move forces with the size equivalent of a brigade by air and sea. The ambition included the procurement of transport aircraft, air refuelling aircraft and naval vessels. The core of the land forces was composed of mechanized and armoured cavalry divisions and brigades as well as of airborne and air cavalry units (Polish MoD 2009).

In 2013, the government noted the importance of counter-surprise attacks. Reinforcing the tactical mobility of the land forces, making the units 'rapidly deployable across the Polish territory, depending on the location of a danger', was hence prioritised. 'Such mobility should be primarily based on the use of helicopters', the government argued. Another prioritised area was the modernisation of the air defence system, 'including the building of a missile defence system aimed to insure more effective national security, as attacks from above are the easiest for the enemy to launch and may reach the Polish territory in a very short period of time'. The third prioritised area focused on computerized combat and support systems, 'including surveillance and command systems, unmanned air warfare, precision-guided weapons and cyber defence measures' (Polish MoD 2013:205). 'The Polish industrial defence capacities [. . .] shall be engaged to the maximum extent in the process of technical modernisation' of the Polish armed forces, the government declared (Polish NSB 2014:34).

Following the Russian war against Ukraine, Minister of National Defence Antoni Macierewicz declared that the number one priority when developing the Polish armed forces was 'the necessity of adequately preparing Poland to defend its own territory'. Hence, establishing the territorial defence forces was one of the measures undertaken. The government also concluded that there was a need to establish a fourth army division for the operational forces. Enhanced capabilities of rocket and tube artillery, new armoured vehicles, new generation MBTs and a new air defence system were the prioritised areas when developing the land forces. In addition, the number of special forces units was to increase, an increased number of coastal missile units and submarines were the key areas when increasing the capacity of the naval forces while new attack helicopters and combat aircraft were the priority of the air force. 'A crucial element of our deterrence force will be played by the Air Force equipped with long-range precision weapons and the 5th generation combat aircraft, the number of which will steadily grow', the government announced (Polish MoD 2017:6 and 51). The government also put emphasis on enhancing Poland's cyber warfare capacity (Polish Ministry of Digital Affairs (MoDA) 2017). Furthermore, the government presented a specific strategic concept for maritime security. Notably, the bulk of the Polish navy, including five submarines, two frigates, one corvette and three attack craft, was not considered 'adequate to the level of threats, challenges, and opportunities generated by the country's maritime security environment'. Responding to 'threats generated by Russia in the Baltic Sea area is not possible with the use of [the] present potential', the government concluded (Polish NSB 2017:32). Improving the capacity for HNS in order to increase the military presence of NATO on its Eastern flank was yet another key area (Polish NSB 2020).

Arguably, the arrangements with the US went further than just being HNS. Several training areas and military complexes were to be at the disposal of the American armed forces. The facilities included six airbases, each with bed-down areas supporting up to 550 personnel, and barracks for altogether 7,200 personnel from a US armoured brigade combat team and US special operations forces (Polish

132 *The empirical exploration*

Table 11.1 Main military resources of Poland.

	2000	2010	2020
Army division	8	4	4
Army brigade (independent)	5	2	2
MBT	1,704	946	606
ACV/APC/IFV	1,438	1,508	1,979
Combat aircraft	267	128	95
Principle surface combatant	1 destroyer, 2 frigates, 4 corvettes, 21 missile & patrol craft	3 frigates, 5 corvettes	2 frigates, 4 corvettes

Source: International Institute for Strategic Studies (2000, 2010, 2020).

Government 2020). Arguably, this represents an additional aspect of the Polish reliance on the US defence industry regarding procurement for the armed forces.

11.5 Ways

'For Poland as a NATO member, any war, irrespective of the scale, would be waged within the Alliance in accordance with the principle that an attack against any NATO member state, including Poland, shall be considered an attack against the whole Alliance', the government declared in 2001 (Polish MoD 2001:15). In 2003, the government emphasised the need to strengthen international cooperation even further. 'Poland shall continue its policy of active engagement in the maintenance of international peace and security on both a regional and global scale', the government pledged. The taking on of the stabilization role in Iraq had, according to the government, enhanced 'Poland's international standing and the due execution of the mission entrusted to us will add to Poland's prestige and image as a responsible and dependable partner on the international scene'. NATO and the bilateral cooperation with especially the US but also with other major member states 'constitute the most important guarantee of external security and stable development of our country', the government declared (Polish NSB 2003:3 and 5).

In 2007, the government announced that NATO was considered as Poland's 'most important form of multilateral cooperation in a political and military dimension of security and a pillar of stability on the European continent, as well as the main ground of transatlantic relations'. The government pledged its support for NATO's engagement in stabilisation missions outside Europe, provided, however, 'that the Alliance maintains a credible potential and is fully capable of collectively defending its member states'. In addition, the government concluded that membership in the EU had a positive effect on Polish security. However, the government hence emphasised the political, economic and social aspects, especially energy security, rather than the military. The government nevertheless pledged to 'make a significant contribution to the development of EU's military and civil capabilities in the field of crisis response', including the creation of and participating in 'the European Rapid Reaction Force' as well as other military task groups. In addition to the US, the bilateral relations were established with Germany, Lithuania and the other Baltic States, as well as with Czechia, Slovakia and Hungary (Polish NSB 2007:10–12).

Notably, two years later, the government claimed that the developments of the EU's CFSP and ESDP were significantly important for Poland. The government announced its intention to continue taking an active part in the developments of the EUBG. The EU 'provides a framework for cooperation in the procurement of military capabilities and in the consolidation of the defence industry. Poland will support and will be actively involved in further development of the European Defence Agency', the government declared. When it comes to the use of force and engagement in international crisis response efforts, contributions to US-, NATO- and EU-led operations were prioritised, but the government did not exclude participation in missions led by the UN or as part of emergency coalitions (Polish MoD 2009:6).[3]

In 2013, the government admitted that it had considered three options: (1) maximum internationalisation, (2) strategic autarchy based on both self-reliance and self-sufficiency, and (3) sustainable internationalisation and autonomy of Poland's security policy. The third option was decided as the way to proceed. 'This option consists in preparing the security system both for seizing the opportunities resulting from international cooperation and for the rational development of capabilities in order to oppose military and non-military threats jointly with allies or individually, if need be', the government argued (Polish MoD 2013:15). At the same time, the government concluded that 'the significance of regional cooperation keeps gaining in importance. For Poland, the key formats in this respect are the Weimar Triangle [that is France, Germany and Poland] and the Visegrad Group' (Polish NSB 2014:9).

On several occasions, Poland has assumed the role of framework nation. In 2010, Poland acted as a framework nation for a battle group established together with Germany, Latvia. Lithuania and Slovakia. In 2013, Poland was responsible for coordinating the Weimar Combat Group, including forces from France and Germany as well as from Poland. In 2016 and 2019, Poland acted as leading nation for the Visegrád Combat Group with Czechia, Hungary and Slovakia, as well as Ukraine in 2016 and Croatia in 2019 (Polish Government 2022).

11.6 Conclusions: Polish strategy

An initial impression may be that Poland has applied an alignment strategy based on multiple-courting. Not only the memberships in NATO and the EU but also in the UN, as well as other organisations and bi- and multilateral arrangements have been described as crucial. Notably, Polish armed forces have participated in several operations led by these organisations. However, we argue that the most prioritised affiliation is the strategic partnership with the US. Polish armed forces contributed, along with American, Australian and British troops, in the US-led invasion of Iraq without a mandate from the UNSC and despite the protests of other NATO and EU members including France and Germany. Moreover, Poland has willingly allowed the US to establish facilities not only for the US ballistic missile defence programme but also for conventional forces on Polish soil. Notably, some of these forces had previously been based in Germany. Consequently, we argue that the Polish alignment strategy foremost is based on *offensive bandwagoning*, that is for profit, with the US.

Regarding the ends, we conclude that the Polish strategy indicates a balance between *survival* and *status*. On the one hand, Poland perceives a clear and present danger from Russia, hence a need to focus on survival. On the other hand, Poland perceives itself as a new but prominent member of the Western community. Notably, the Polish government had strived for recognition rather than influence. When it comes to means, we argue that Poland over time has focused on *national defence*. Despite impressive contributions to several, differently led international military operations, the focus has remained on the defence of Poland. Obviously, the contributions to several different contexts regarding the use of force, as well as several bi- and multilateral arrangements regarding the development of military power, indicate a preference for a *multilateral approach* when it comes to the ways. However, the Polish government has continuously expressed its preparedness for a *unilateral approach* if deemed necessary. Arguably, this is a central part of the Polish deterrence strategy.

Table 11.2 Polish strategy.

Alignment strategy	Military strategy		
	Ends	Means	Ways
Offensive bandwagoning	Survival/status	National defence	Multilateral and unilateral approach

We agree with Krzysztof Miszczak when claiming that the development of ESDP does not meet the Polish demands of 'hard guarantees' (Miszczak 2007:31). However, we do agree with Hrvoje Ćiković when including energy security in the elaborations. Obviously, the Polish government has stressed this aspect over time. Nevertheless, we argue that Poland has strived to address this challenge through EU rather than through NATO channels. Moreover, the Polish government has separated this aspect of grand strategy from military considerations (Ćiković 2008). This is in line with the findings of Laura Chappell. In addition, she shed interesting light on the divergence between Polish and EU policies regarding Russia as well as the Polish Atlanticism. Arguably, her claims that this Atlanticism has focused on the US rather than on NATO support our conclusion regarding the bandwagoning alignment strategy (Chappell 2010; see also Kupiecki 2019; Sliwa 2019; Smura 2019).

We appreciate Krzysztof Załęski illuminating the Polish air force's activism within the framework of the US-led operations in Iraq and Afghanistan. Arguably, this also supports our conclusions (Załęski 2012). This goes for Steven Dubriske's findings on the Polish–American agreement regarding missile defence as well (Dubriske 2012). We find Laura Chappell's conclusions regarding the Polish strategic culture interesting and agree with her when she claims that 'Poland's experience of defeat [. . .] has shaped the country's security and defence policy' (Chappell 2012:172; see also Doeser 2018). As previously touched upon, we also agree with her findings that Poland tends to be enthusiastic regarding Atlanticism but initially rather sceptical regarding the CSDP, since it did not fully encompass Poland's more traditional threat perceptions. Consequently, we agree with Justyna Zajac when she concludes that 'the main goal of Poland's security policy has been to provide protection against Russia' (Zajac 2016:185). We find her arguments on the explanation of this position being Poland's negative historical experience and geographical location convincing. Obviously, we agree with her conclusion, defining the Polish strategy as bandwagoning with the US.

In this context, the findings of Joseph Bell and Ryan Hendrickson provide a more nuanced image by focusing on Poland's position during NATO's Operation Unified Protector over Libya in 2011 (Bell and Hendrickson 2012). Potentially, Poland's and the other Visegrád countries' freeriding in this specific operation can be explained by the American reluctance to lead. As indicated by Juraj Marušiak, the Visegrád group is far from homogeneous when it comes to the use of military force (Marušiak 2015). This is in line with our findings regarding the difference between creating and using military force. We agree with Mark Yaniszewski when arguing that going too far with the reduction of the military means in the transformation process could threaten the very viability of the armed forces. Clearly, the Polish government has had this in mind (Yaniszewski 2002). We find support from Tomasz Paszewski regarding our analysis on the ways, balancing between multi- and unilateralism. Obviously, we agree with his conclusion that 'Poland's efforts to enhance its national-defence capabilities are a natural complement to the strengthening of NATO's defences' (Paszewski 2016:127; see also Epstein 2016).

Fredrik Doeser and Joakim Eidenfalk (2019) provide interesting insights in this regard. Obviously, we agree when they argue that Poland recently has embarked on a more cautious line and potentially has broken with its previous policy of active participation in US-led operations outside Europe. We also agree with their observation regarding the importance for the Polish strategy having not only American but also NATO troops stationed on Polish territory. However, we do not consider this aspect indicating a shift towards a *balance of power* alignment strategy. At least not for the moment.

Notes

1 We would like to express our gratitude to Colonel Ulf Jonsson, Embassy of Sweden, Warzaw, for his support.
2 Jerzy Buzek of the Solidarity Electoral Action (AWS) served as prime minister 31 October 1997–19 October 2001. Leszek Miller of the Democratic Left Alliance (SLD) served as prime minister 19 October 2001–2 May 2004. Marek Belka of the SLD served as prime minister 2 May 2004–31 October 2005. Kazimierz Marcinkiewicz of the Law and

Justice (PiS) party served as prime minister 31 October 2005–14 July 2006. Jarosław Kaczyński of the PiS served as prime minister 14 July 2006–16 November 2007. Donald Tusk of the Civic Platform (PO) served as prime minister 16 November 2007–22 September 2014. Ewa Kopacz of the PO served as prime minister 22 September 2014–16 November 2015. Beata Szydło of the PiS served as prime minister 16 November 2015–11 December 2017. As of 1 January 2022, Mateusz Morawiecki of the PiS has served as prime minister since 11 December 2017. See www.gov.pl/web/diplomacy/basic-information.
3 Notably, all services have participated in NATO-led operations. Polish military have also contributed to several EU-led missions, that is Concordia, Althea, EUFOR Tchad/RCA, EUFOR RD Congo, and EU Training Mission (EUTM) Mali, as well as to UN-led operations, that is UN Disengagement Observer Force (UNDOF), UN Interim Force in Lebanon (UNIFIL) and UN Mission in the Central African Republic and Chad (MINURCAT). Moreover, Polish armed forces have participated in the US-led operations Enduring Freedom in Afghanistan and Iraqi Freedom (Polish MoD 2013:50).

Bibliography

Banaszkiewicz-Zygmunt, Edyta and Krzysztof Olendzki (eds) (2000). *Poland: An Encyclopedic Guide*. Warsaw: Polish Scientific Publishers.
Bell, Joseph and Ryan Hendrickson (2012). 'NATO's Visegrad Allies and the Bombing of Qaddafi: The Consequence of Alliance Free-Riders' *Journal of Slavic Military Studies* Volume 25, Issue 2.
Chappell, Laura (2010). 'Poland in Transition: Implications for a European Security and Defence Policy' *Contemporary Security Policy* Volume 31, Issue 2.
――― (2012). *Germany, Poland and the Common Security and Defence Policy: Converging Security and Defence Perspectives in an Enlarged EU*. Basingstoke: Palgrave Macmillan.
Ćiković, Hrvoje (2008). 'The Analysis of the Influence of Energy Security on Foreign and Security Policy of the Republic of Poland' *Croatian International Relations Review* Volume 14, Issue 52/53.
Doeser, Fredrik (2018). 'Historical Experiences, Strategic Culture, and Strategic Behavior: Poland in the Anti-ISIS Coalition' *Defence Studies* Volume 18, Issue 4.
Doeser, Fredrik and Joakim Eidenfalk (2019). 'Using Strategic Culture to Understand Participation in Expeditionary Operations: Australia, Poland, and the Coalition against the Islamic State' *Contemporary Security Policy* Volume 40, Issue 1.
Dubriske, Steven (2012). 'Ballistic Missile Defense in Poland: Did the Costs Outweigh the Benefits?' *Connections* Volume 12, Issue 2.
Epstein, Rachel (2016). 'When Legacies Meet Policies: NATO and the Refashioning of Polish Military Tradition' *East European Politics and Societies* Volume 20, Issue 2.
International Institute for Strategic Studies (IISS) (2000). *The Military Balance 2000–2001*. Oxford: Oxford University Press.
――― (2010). *The Military Balance 2010*. London: Routledge.
――― (2020). *The Military Balance 2020*. London: Routledge.
Kupiecki, Robert (2019). *Poland and NATO after the Cold War*. Warsaw: Polish Institute of International Affairs.
Liddell-Hart, Basil (1997). *History of the Second World War*. London: Papermac.
Marušiak, Juraj (2015). 'Russia and the Visegrad Group: More Than a Foreign Policy Issue' *International Issues & Slovak Foreign Policy Affairs* Volume 24, Issue 1–2.
Miszczak, Krzysztof (2007). 'Poland and the Development of the European Security and Defence Policy' *Polish Quarterly of International Affairs* Volume 16, Issue 3.

Paszewski, Tomasz (2016). 'Can Poland Defend Itself?' *Survival* Volume 58, Issue 2.
Polish Government (2020). *Agreement on Enhanced Defense Cooperation with the USA*.
——— (2022). *Poland's Commitment in the Implementation of the EU Common Security and Defence Policy: Military Dimension*.
Polish Ministry of Defence (MoD) (2001). *White Paper on Defence*.
——— (2009). *National Defence Strategy*.
——— (2013). *White Paper on Defence*.
——— (2017). *National Defence Concept*.
Polish Ministry of Digital Affairs (MoDA) (2017). *National Cyber Security Strategy*.
Polish Ministry of Foreign Affairs (MoFA) (2017). *Foreign Policy Strategy*.
Polish National Security Bureau (NSB) (2003). *National Security Strategy*.
——— (2007). *National Security Strategy*.
——— (2014). *National Security Strategy*.
——— (2017). *Strategic Concept for Maritime Security*.
——— (2020). *National Security Strategy*.
Sanford, George (1999). *Poland: The Conquest of History*. Amsterdam: Overseas Publishers Association.
Sliwa, Zdzislaw (2019). 'Poland: NATO's East European Frontline Nation' in Nora Vanaga and Toms Rostoks (eds). *Deterring Russia in Europe: Defence Strategies for Neighbouring States*. Abingdon: Routledge.
Smura, Tomasz (2019). 'Relations between the United States and Poland: From Enemy to the Main Security Guarantor' in Anna Péczeli (ed). *The Relations of Central European Countries with the United States*. Budapest: Dialóg Campus.
Yaniszewski, Mark (2002). 'Post-Communist Civil-Military Reform in Poland and Hungary: Progress and Problems' *Armed Forces & Society* Volume 28, Issue 3.
Zajac, Justyna (2016). *Poland's Security Policy: The West, Russia, and the Changing International Order*. Basingstoke: Palgrave Macmillan.
Załęski, Krzysztof (2012). 'The Polish Air Force Operations in Air Policing over the Baltic States and the Future of the Mission' *Baltic Security & Defence Review* Volume 14, Issue 2.

12 The strategy of Romania[1]

The primary sources analysed in this chapter consist of the NSSs of 2001 and 2007, the military strategies of 2000, 2016 and 2021, the NDSs of 2015 and 2020, the DWPs of 2016 and 2017 and the policy paper on Romanian defence of 2013.[2]

12.1 Historical background

Moldavia was established as an independent principality in the mid-fourteenth century. In the early sixteenth century, the principality became a vassal state of the Ottoman Empire. From the late sixteenth to early eighteenth century, the principality of Transylvania was a semi-independent state, ruled primarily by Hungarian princes. In 1330, Wallachia was established as an independent principality. During the fifteenth century, it became a vassal state of the Ottoman Empire. With the exception of brief periods of Russian occupation between 1768 and 1854, this vassalship lasted until the nineteenth century. The two principalities of Transylvania and Wallachia were unified in 1859 and adopted the name Romania in 1862. Following the Treaty of Berlin in 1878, Romania was recognized as an independent state. Transylvania was transferred to Romania after the defeat of Austria-Hungary in WWI. In 1940, parts of Transylvania was transferred to Hungary after arbitration of Nazi-Germany and Italy. Concurrently, the USSR annexed Bessarabia and some other parts of northern Romania. Shortly afterwards, Romania joined the Axis powers. In 1941, Romania took part in the German invasion of the USSR. However, as the fortunes of war changed, Romania switched sides. The fact that Romanian forces participated in the Soviet invasions of Hungary, Slovakia and Germany did not prevent the USSR from invading and occupying the country. The monarchy came to last until 1947, when the people's republic of Romania was announced (see, for example, Bachman 1991; Sanborne 1996; Liddell-Hart 1997).

12.2 Strategic environment

In the NSS issued in 2001, the Romanian government expressed a perceived vulnerability regarding the country's security status. 'This situation is the direct result of the cumulative pressure, over time, of a number of multiple factors',

DOI: 10.4324/9781003298052-14

the government explained and mentioned political, economic, financial, social, cultural, biological, religious, demographic and military factors. These factors had, the government argued, 'influenced the safe environment of the state and citizens, leading to the weakening of the moral, material standing, as well as that of spiritual values, on which the civilization of our national identity is based'. The government observed that the 'risks of the emergence of a traditional military confrontation on the European continent have seriously diminished'. Hence, the government concluded that 'Romania is not confronted and will not be, in the near future, with major threats of classic military types, against its national security' (Romanian President 2001:2, 7 and 14). The government identified four regional risks:

- strategic imbalance in military capability within Romania's area of strategic interest;
- military conflicts and tensions which could extend;
- standing economical-social shortcomings directly affecting military capability and depreciating the authority of national leadership institutions of our state;
- the possibility of disrupting financial, information, energy, communications and telecommunications of the states' systems, and the political-military rivalries between them.

(Romanian MoD 2000:21)

In 2007, the government argued that the main confrontation lines in the security environment were caused by the fundamentally different values between democracy and totalitarianism and by identity differences of ethnic, religious, cultural and/or ideological natures. International religious-extremist-driven terrorism was considered as a trigger, while the access to strategic resources was considered as a driving force in this regard. In addition to terrorism, the proliferation of WMD, regional conflicts, organised crime and climate change were considered as the main challenges. Regarding the regional or separatist conflicts, the government especially mentioned the Dniester region in Moldova, Abkhazia and South Ossetia in Georgia, Nagorno-Karabakh in Azerbaijan and Chechnya in Russia. 'The likelihood of a wide-ranging military conflict is slim', the government nevertheless concluded (Romanian President 2007:10). In 2015, the government concluded that, at the global level,

> the security environment is undertaking an ongoing transformation process, which reflects mainly upon highlighting interdependences and unpredictability within international relations system and the difficulty to delimitate classical risks and threats from the asymmetric and hybrid ones.
> (Romanian President 2015:11)

The government stressed that Romania's geographical position at the crossroads of some areas with a high security risks and tensions, made it necessary to pay specific attention not only to the developments in the Middle East and in Northern Africa but also to those in the Black Sea Region, not least following Russia's annexation of Crimea. 'The intensification of migration from the conflict areas or from areas with a poor economic situation has generated challenges [...] to manage the flow of illegal immigrants', the government admitted (Romanian President 2015:13). Clearly, Russia's efforts to strengthening its great power status and the consequences thereof for countries such as Ukraine, Moldova and Georgia worried the government. Moreover, cyber attacks 'are at an all-time high and represent an important category of threats with a global reach which are difficult to identify and fight', the government observed (Romanian Parliament 2016:13). One year later, the government argued that Europe

> is facing major challenges, with unpredictable consequences on its security. On the one hand, it is necessary to identify a coherent solution to respond to [the] Russian Federation and hybrid threats' actions, which are more and more persistent at the eastern border, and on the other hand it is necessary to manage the wave of migrants from areas of instability in the Middle East and North Africa.
>
> (Romanian MoD 2016:7)

In 2020, the government predicted that the international security architecture would be strongly determined and redefined by state actors having global interests. Especially China was believed to be increasing its influence at the expense of the US. Naturally, the COVID-19 pandemic and its economic consequences were elaborated, but the intensified arms race also worried the government. 'The development of new and increasingly powerful military resources and their expansion into new regions such as the Arctic and Antarctica, the cosmic space or cyberspace, leads to increased insecurity and risks including those occurring accidentally', the government warned (Romanian President 2020:21).

12.3 Ends

In 2001, the Romanian government listed a number of national interests to be safeguarded. While the first interest concerned the maintenance of 'the integrity, unity, sovereignty and independence' of the Romanian state, the second focused on guaranteeing 'the fundamental and democratic freedoms, and ensuring the welfare, security and safety' of Romania's citizens. The economic and social development of the country were fundamental in the elaborations. Consequently, an intense 'reduction of the wide gaps separating Romania from the developed European countries' was a key objective. Meeting the conditions for Romania's integration as a NATO and EU member was another central aim. 'Romania must become a component with full obligations and rights of the two organizations, the only ones capable of guaranteeing its independence and sovereignty and enable an economic, political and social development similar to that of the democratic countries', the government argued. Optimising the national defence capability in order to keep up with the NATO standards was hence considered not only a fundamental objective but a necessity. At the same time, asserting 'the national identity and pursuing it as a democratic value, making best use of and developing the national cultural heritage and the creative abilities of the Romanian people' were declared national interests. Consequently, diversifying and strengthening ties with Romanians living outside the national borders of Romania was considered a central objective (Romanian President 2001:4).

The government also provided a list of some national military objectives. The first of these objectives focused on preventing conflicts and managing crisis that otherwise could affect the military security of Romania directly. Another objective focused on preventing, deterring and defeating 'any possible armed aggression against Romania' and a third on the 'gradual integration into NATO military structures' (Romanian MoD 2000:7). In 2007, Romanian President Traian Băsescu declared his ambition to improve Romania's status and, in order to,

> achieve its rightful interests, in its position as an integral part of the Euro-Atlantic civilization and an active participant in the process of building the new Europe, Romania: promotes, protects, and defends democracy; observes the fundamental human rights and liberties; takes actions that comply with inter-national law, in order to speed up its economic and social modernization and development; acts to ensure the European living standards; and asserts its national identity.
>
> (Romanian President 2007:3)

Actively supporting Romanian communities living in 'their historical area' as well as Romanians all over the world in order to preserve their national and cultural identity was also described as a core objective (Romanian President 2007:18). Romania 'is a state that ensures the security of its citizens wherever they are', Romanian President Klaus Iohannis made clear in 2015. Romania intended to 'reinforce its strategic credibility, will be known for its predictable and constant stance on defining its foreign, security and defence policies, as well as on strengthening democracy and the rule of law', he continued (Romanian President 2015:3). Notably, the formulation of the national security interests was slightly adjusted:

- guaranteeing the state's national character, sovereignty, independence, unity and indivisibility;
- defending the country's territorial integrity and inalienability;
- defending and consolidating constitutional democracy and the rule of law;
- protecting fundamental rights and liberties of all citizens and guaranteeing their safety.

(Romanian President 2015:8)

Strengthening Romania's profile within NATO and the EU through operational contributions and upholding Moldavia's 'European aspirations' were hence

142 *The empirical exploration*

declared as key objectives (Romanian President 2015:10). In the military strategy issued in 2021, the missions of the Romanian Armed Forces were declared to be

- guaranteeing the sovereignty, independence and unity of the state, the territorial integrity of the country and constitutional democracy;
- contributing to collective defence in military alliance systems;
- participating in actions for maintaining or restoring peace;
- contributing to Romania's security in peacetime; and
- supporting recognised authorities during civil emergencies.

(Romanian MoD 2021:11)

12.4 Means

In 2001, the Romanian government admitted the necessity of continuing the transformation of the armed forces in accordance with NATO and EU standards. The aim of the reforms was not only 'to develop a credible, modern and effective defence capability' but also to build up 'Romania's status as a security provider, by continuing and improving its contribution to regional stability'. By downsizing, restructuring and streamlining the forces, a proper dimension of the command element was to be achieved. Regulating the retirement of the manpower made redundant from not only the armed forces but also from the defence industry was hence a key priority (Romanian President 2001:23). The government announced that the future military force of Romania was to consist of 112,000 military personnel and 28,000 civilians. The number of professional military personnel was hence to increase from 47 to 71 percent. The operational structure of the land forces was to include eight combat brigades, four combat support brigades and two logistic brigades, while the reserve structure was to include ten combat brigades, five combat support brigades and two logistic brigades. The active units of the air force were to consist of two air division commands, four airbases and two air defence brigades, while the reserve units were to consist of two additional airbases and three airfields. Notably, the government did not provide any detailed information regarding the naval forces. The government also declared its ambition enhancing the bilateral cooperation with not only other NATO members but with NATO candidate countries and other partners as well. The ambition was to take active part in (1) the Multinational Peace Force–South-Eastern Europe (MPF-SEE), together with Albania, Bulgaria, Greece, Italy Macedonia and Turkey, (2) the Multinational Stand-by High Readiness Brigade (SHIRBRIG), and (3) the Black Sea Naval Cooperation Group (BLACKSEAFOR) with Bulgaria, Georgia, the Russian Federation, Turkey and Ukraine (Romanian MoD 2000). In 2007, the government announced that

> Romania aims to create a modern and professional Army that has the proper number of troops and appropriate equipment, with mobile and multifunctional expeditionary forces that are swiftly deployable, flexible and effective.

They have to be able to provide a reliable defence of the national territory, to fulfil its commitments to collective defence, and to take part in international operations.

(Romanian President 2007:50)

In 2013, the government announced that the MPF-SEE, also known as the South Eastern Europe Brigade (SEEBRIG), was to include Bosnia-Herzegovina and Ukraine. In addition, the Romanian-Hungarian Joint Peacekeeping Battalion was declared operational. The government also informed that the deployment of the American AEGIS Ashore system to the Deveselu site in Romania would take place in 2015, hence being a crucial part of NATO's missile defence capability. The government compared this development with other Romanian contributions to NATO that, in addition to land forces deployed to Afghanistan, the Balkans and Iraq, included four MiG-21 combat aircraft to the air policing mission in the Baltic States in 2007, the frigates deployed to Operation Active Endeavour in the Mediterranean Sea from 2005 through 2011 and the frigate deployed to Operation Unified Protector in 2011. At its peak, Romania's deployment to Afghanistan included 1,800 troops in 2012, the government recalled. The government also mentioned the Romanian contributions to the EU-led operation Althea in Bosnia-Herzegovina and the training mission to Mali. Notably, each of these contributions consisted of fewer than 40 military personnel. However, the contributions to Operation Atalanta on the Indian Ocean included a frigate (Romanian MoD 2013). When developing the armed forces, the government gave priority to maintain and develop 'response capabilities to carry out actions to deter and counter a possible armed aggression against Romania, until the intervention of allied forces' and to ensure 'the capabilities required by the Enhanced NATO Joint Response Force and by the EU Battle Groups'. The government announced its ambition of reaching the goal of allocating a minimum of 2 percent of the GDP for defence purposes by 2017 but warned about continued budgetary constraints. Regarding the new acquisitions and modernisations, wheeled and tracked combat vehicles, artillery systems, and division and brigade-type command posts were prioritised regarding the needs of the army. Both combat and airlift aircraft were mentioned regarding upgrading the air force, while type T22R frigates, multifunctional corvettes, as well as missile boats and other naval platforms, were mentioned regarding the navy (Romanian Parliament 2016:32–33). The government announced that the new organisation of the army was to include two division headquarters, eight combat brigades, two combat support brigades, and one intelligence, surveillance, and reconnaissance brigade. The new force structure of the air force was to include five airbases and one air defence brigade, while the naval resources were to be organised in one maritime fleet, one frigate flotilla and one river flotilla (Romanian MoD 2016). Of these resources, a division-level operational group land force, adequately supported by both air and naval forces, was to constitute the bulk of Romania's contribution to the collective defence efforts of both NATO and the EU. 'Romania provides the same force package to both organizations', the government clarified (Romanian MoD 2017:34).

Table 12.1 Main military resources of Romania.

	2000	2010	2020
Army corps	7	0	0
Army division HQ	0	2	2
Army brigade (independent)	0	7	10
MBT	1,253	299	400
ACV/APC/IFV	1,796	1,095	895
Combat aircraft	323	49	56
Principle surface combatant	1 destroyer, 6 frigates, 38 (offshore) missile and patrol craft	3 frigates, 4 corvettes, 11 (offshore) missile and patrol craft	3 destroyers, 4 corvettes, 14 (offshore) missile and patrol craft

Source: International Institute for Strategic Studies (2000, 2010, 2020).

In 2020, the government announced its intentions establishing several installations on Romanian territory in order to enable and strengthen NATO enhanced forward presence. At the same time, the government made clear that Romania 'must have its own resources, forces and means, in line with its interests, profile and potential' (Romanian President 2020:14). In the military strategy of 2021, the government presented the structure of the armed forces. The land forces were to organise two division headquarters with the related combat and logistic support structures, eight manoeuvring brigades, two combat support brigades, and one research, surveillance and reconnaissance brigade. The air force was to operate five airbases, while the navy was to organise one maritime fleet and a river flotilla. However, the number of either combat aircraft or combat ships was not specified. In addition, the bulk of the special forces command included two special operations battalions, one commando battalion and one maritime special operations battalion (Romanian MoD 2021).

12.5 Ways

The NSS released in 2001 focused on making Romania's armed forces 'assigned to participate in EU missions, in the framework of the European security and defence policy, operational, as well as in NATO/UN missions and of subregional forums/undertakings'. Strengthening the enhanced strategic partnership with the US, developing privileged bilateral relationships with NATO and EU member states, as well as building up bi- and multilateral relationships with neighbouring countries were hence on the agenda. The 'privileged relationships with the Republic of Moldavia' was especially to be developed (Romanian President 2001:23–24). 'Romanian Armed Forces are and will be capable of repelling and if necessary, of defeating a possible military aggression, within the limits of our capabilities', the government declared. The military were also 'prepared to promote the national interests' in other contexts, the government announced. Several different concepts for developing the armed forces were presented. One such concept, gradual and flexible response, focused on the use of military force for national defence. The concept included the peacetime deployment of forces both in 'the border areas and in the depth of the national territory' in order to 'provide surveillance and early warning capability'. The concept aimed at 'preventing strategic surprise, to discover and monitor any indication related to the emergence and development of crisis and conflict situations or the danger of breaking out an armed aggression against Romania'. In case of a large-scale military aggression against Romania or against one or several of Romania's allies and partners, Romanian armed forces were to 'be capable of participating in joined and combined operations in compliance with the collective defence principles and with the provisions of the new NATO Strategic Concept'. In addition to the strategic partnership with the US, the government stressed the importance of the special partnerships developed with the UK, Germany, France and Italy (Romanian MoD 2000:2 and 8–9).

In 2007, the government advocated for proactively countering risks and threats in a timely manner by actively contributing to international crisis prevention and management. NATO, the EU, the UN and the OSCE were hence considered as being of 'significant importance'. The government also mention pre-emptive measures in order to effectively counter international terrorism as well as the proliferation of WMD (Romanian President 2007:21). The strategic partnership with the US and the NATO and EU memberships represent 'the key pillars' of Romania's security policy, the Romanian president declared in 2015 (Romanian President 2015:4). The Parliament concluded that the main:

> guarantee for Romania's security remains NATO, whose policies and capabilities are based on the principles of allied solidarity and indivisibility, on maintaining a strong transatlantic relationship, on the firm U.S. commitments in Europe, and on the strengthening of Alliance partnerships.
> (Romanian Parliament (2016:11)

In the military strategy issued in 2021, Minister of National Defence Mr. Nicolae-Ionel Ciucă stressed that, in addition to NATO and the EU,

> Romania will continue to rely on its strategic partnership with the United States, which provides the optimal framework for strengthening relations between the two countries' armed forces in order to substantiate Romania's strategic position on the Black Sea.
> (Romanian MoD 2021:2)

12.6 Conclusions: Romanian strategy

Although the Romanian government was already very supportive towards the US before Romania's entrance into NATO and the EU, despite the special strategic partnership between Romania and the US, we do not consider bandwagoning as the most appropriate label when it comes to the country's alignment strategy. Rather we argue that Romania adopted a *multiple-courting*, that is hedging, alignment strategy. Even though memberships in NATO and in the EU, and the partnership with the US were a prioritized end, Romania strived for establishing other strategic arrangements as well, including with the UN. Notably, before the war in Ukraine, even Russia was considered a potential partner.

Regarding the ends of the military strategy, we conclude that *status* is the most appropriate label at the aggregated level. The government has expressed objectives indicating considerations on influence as well, but we argue that gaining recognition within NATO and the EU, especially from their key members, as an important and credible member/partner was prioritised. Potentially, once the wanted status was achieved, Romania might very well shift focus in order to increase its influence. Although the government continuously has emphasised not only the sovereignty and integrity of the Romanian state, as well as the cultural and historical identity of the Romanian nation, survival has not been the dominant aspect.

When it comes to means, the Romanian government has not always explicitly declared its priorities. Clearly, the armed forces have been downsized but the capacity for national defence has nevertheless remained at a high level. Taking all contributions to international military operations into account, we conclude that *expeditionary warfare* has been the focus. Notably, units from all three services have been deployed abroad, including combat aircraft and frigates. At its peak, Romania had about 1,800 troops deployed just in Afghanistan. Moreover, the armed forces have been deployed in different organisational contexts, hence putting deeds behind the multiple-courting approach. Finally, regarding the ways, we find that these deployments also make the *multilateral approach* trustworthy. Although we observe a tendency towards a more bilateral approach focusing on the US, the ambition to participate in both NATO and the EU still is stronger in this regard. Arguably, this goes hand in hand with the end being recognised with a certain status.

Table 12.2 Romanian strategy.

Alignment strategy	*Military strategy*		
	Ends	*Means*	*Ways*
Multiple-courting	Status	Expeditionary warfare	Multilateral approach

We agree with Nicolae-Stefan Ciocoiu when, on the one hand, he observed the considerable reduction of army brigades and aircraft squadrons while one the other hand noting that the number of frigates increased. We also agree that this adjustment took place in order to make Romania more interoperable regarding NATO's Operational Concept and Force Structure (Ciocoiu 2004). Obviously, we find Mihail Ionescu supportive when shedding light on the Romanian–Russian bilateral treaty for good neighbourliness and friendship relations. Arguably, this is in line with our conclusions regarding multiple-courting (Ionescu 2005). We find the conclusion of Liviu Muresan and Marian Zulean that 'Romanian defense and security policy has developed over time, from a nationalistic autarchy to a regional player with global aspirations, within the framework of the NATO team' in line with our findings (Muresan and Zulean 2008:174). We also find that Ștefan Dănilă and Avram-Florian Iancu support our conclusions when emphasising Romania's quest for being perceived as a credible partner and as a real contributor to Euro-Atlantic security (Dănilă and Iancu 2014; see also Csernatoni 2014).

Arguably, the findings of Iulia-Sabina Joja regarding national interests, political orientation and national identity are in line with our own findings in this regard (Joja 2015). We also find support in the conclusions presented by Florentin-Gabriel Giuvara and Marius Șerbeszki regarding the Romanian partnership with the US. We agree with their argument that aspects of Romania's economic security and potential foreign investments also need to be taken into account in this regard (Giuvara and Șerbeszki 2015). We also agree with Carmen Sorina Rîjnoveanu when arguing that Romania is unable to provide sufficient deterrence vis-à-vis Russia solely by its own resources, hence making the strategic partnership with the US and the NATO engagement the core of Romania's doctrinaire approach (Rîjnoveanu 2017; see also Rîjnoveanu 2019). Naturally, we agree with Siemon Wezeman and Alexandra Kuimova when observing that the 'Romanian armed forces still operate equipment largely acquired before the end of the cold war' and that 'by the mid-2000s much of it had become outdated'. We find their detailed elaboration on potential procurements useful (Wezeman and Kuimova 2018:12). Finally, we find the conclusions presented by Mihai Vladimir Zodian being in line with ours regarding the alignment strategy based on multiple-courting (Zodian 2019).

Notes

1 We would like to express our gratitude to Colonel Ciprian Antonescu, Embassy of Romania, Oslo, and Lieutenant Colonel Jörgen Marqardsen, Embassy of Sweden, Bucarest, for their support.
2 Mugur Isărescu served as independent prime minister 22 December 1999–28 December 2000. Adrian Năstase of the Social Democratic Party (PSD) served as prime minister 28 December 2000–21 December 2004. Eugen Bejinariu of the PSD served as prime minister 21–28 December 2004. Călin Popescu-Tăriceanu of the National Liberal Party (PNL) served as prime minister 28 December 2004–22 December 2008. Emil Boc of the Democratic Liberal Party (PDL) served as prime minister 22 December 2008–6 February 2012. Cătălin Predoiu served as interim prime minister 6–9 February 2012. Mihai Răzvan Ungureanu served as interim prime minister 9 February–7 May 2012. Victor Ponta of the PSD served as prime minister 7 May 2012–5 November 2015 with the exceptions of 22 June–9 July 2015 and 29 July–10 August 2015, when Gabriel Oprea of the National Union for the Progress of Romania (UNPR) served as prime minister. Sorin Cîmpeanu of the Alliance of Liberals and Democrats (ALD) served as prime minister 5–17 November 2015. Dacian Cioloș served as independent prime minister 17 November 2015–4 January 2017. Sorin Grindeanu of the PSD served as prime minister 4 January–29 June 2017. Mihai Tudose of the PSD served as prime minister 29 June

148 *The empirical exploration*

2017–16 January 2018. Mihai Fifor of PSD served as prime minister 16–29 January 2018. Viorica Dăncilă of the PSD served as prime minister 29 January 2018–4 November 2019. Ludovic Orban of the PNL served as prime minister 4 November 2019–7 December 2020. Florin Cîțu of the PNL served as prime minister 23 December 2020–25 November 2021. Nicolae Ciucă of the PNL served as prime minister 7–23 December 2020 and has, as of 1 January 2022, served as prime minister since 25 November 2021. See www.gov.ro/en.

Bibliography

Bachman, Ronald (ed) (1991). *Romania: A Country Study*. Washington, DC: Library of Congress.
Ciocoiu, Nicolae-Stefan (2004). *Romanian Armed Forces Transformation Process: The Core Issue of the National Military Strategy towards NATO Integration*. Carlisle, PA: U.S. Army War College.
Csernatoni, Raluca (2014). *Romania's Euro-Atlantic Security Profile Post-Cold War: Transitional Security Habitus and the Praxis of Romania's Security Field*. Budapest: Central European University.
Dănilă, Ștefan and Avram-Florian Iancu (2014). 'Romanian Armed Forces a Decade after NATO Accession: An Institutional Perspective' *Strategic Impact* Volume 51, Issue 2.
Giuvara, Florentin-Gabriel and Marius Șerbeszki (2015). 'Missile Defence in Romania: Implications for Security Policy' *Impact Strategic* Volume 54, Issue 1.
International Institute for Strategic Studies (IISS) (2000). *The Military Balance 2000–2001*. Oxford: Oxford University Press.
——— (2010). *The Military Balance 2010*. London: Routledge.
——— (2020). *The Military Balance 2020*. London: Routledge.
Ionescu, Mihail (2005). 'Romania's Position towards the Evolution of the Transatlantic Link after 11 September 2001' in Tom Lansford and Blagovest Tashev (eds). *Old Europe, New Europe and the US: Renegotiating Transatlantic Security in the Post 9/11 Era* Abingdon: Routledge.
Joja, Iulia-Sabina (2015). 'Reflections on Romania's Role Conception in National Strategic Documents 1990–2014: An Evolving Security Understanding' *Europolity* Volume 9, Issue 1.
Liddell-Hart, Basil (1997). *History of the Second World War*. London: Papermac.
Muresan, Liviu and Marian Zulean (2008). '"From National, Through Regional, to Universal . . ." Security and Defense Reform in Romania' in Thomas Bruneau and Harold Trinkunas (eds). *Global Politics of Defense Reform*. Basingstoke: Palgrave Macmillan.
Rîjnoveanu, Carmen Sorina (2017). 'The Dynamics of Deterrence in the Black Sea Area: The Challenges to Romania's Security Agenda' *Monitor Strategic* Issue 3–4.
——— (2019). 'Romania's Approach to Deterrence' in Nora Vanaga and Toms Rostoks (eds). *Deterring Russia in Europe: Defence Strategies for Neighbouring States*. Abingdon: Routledge.
Romanian Ministry of Defence (MoD) (2000). *Military Strategy*.
——— (2013). *Romanian Defence*.
——— (2016). *Military Strategy*.
——— (2017). *White Paper on Defense*.
——— (2021). *Military Strategy*.
Romanian Parliament (2016). *White Paper on Defense*.
Romanian President (2001). *National Security Strategy*.

——— (2007). *National Security Strategy*.
——— (2015). *National Defense Strategy*.
——— (2020). *National Defense Strategy*.
Sanborne, Mark (1996). *Romania*. New York: Facts On File Inc.
Wezeman, Siemon and Alexandra Kuimova (2018). 'Bulgaria and Black Sea Security', *SIPRI Background Paper*, December.
Zodian, Mihai Vladimir (2019). 'Relations between the United States and Romania: Changing Dynamics?' in Anna Péczeli (ed). *The Relations of Central European Countries with the United States*. Budapest: Dialóg Campus.

13 The strategy of Slovakia[1]

The primary sources analysed in this chapter consist of the NSSs of 2001 and 2005, the NDSs of 2001, 2005 and 2017, the military strategy of 2001, the military doctrine of 2003, the DWPs of 2013 and 2016 and the foreign and European policy of 2020. Moreover, the Cyber Security Concept of 2015 and the action plan for its implementation as well as the concept for the fight against hybrid threats of 2018, along with the proposal for the preparation of a new security strategy of 2020, have been studied.[2]

13.1 Historical background

Following the dissolution of the Moravian Empire in the early eleventh century, the territory of present Slovakia was integrated into the Hungarian Kingdom, to be known as Upper Hungary. Following the Ottoman conquest of Hungary in 1526, Upper Hungary became part of the Habsburg Monarchy. However, a Hungarian-led rebellion against the Habsburgs in the 1680s led to the short-lived establishment of the Principality of Upper Hungary as a vassal state to the Ottoman Empire. In 1918, Slovakia was part of the newly independent state, Czechoslovakia. Shortly before WWII, a German-led arbitration led to Hungary gaining some territory on behalf of the Czechoslovakians. Following the final annexation of the Czech parts of Czechoslovakia by Nazi-Germany and Hungary in March 1939, the Slovakian part formed a nominally sovereign but in reality a German-controlled puppet state, that is the Slovak Republic. The Slovaks took part on the side of Nazi-Germany against Poland as well as against the USSR. During the final phase of WWII, a Slovakian uprising took place, leading to German occupation. This was soon followed by a Soviet invasion and occupation. After WWII, the USSR oversaw the establishment of the third Czechoslovak republic. The Slovakians thereafter shared the same destiny as the Czechs until the peaceful dissolution of Czechoslovakia on 31 December 1992 (see, for example, Wallace 1977; Kirschbaum 1995; Gorys 1996; Liddell-Hart 1997).

13.2 Strategic environment

In 2001, the Slovakian National Council (NC) concluded that there was an increased number of security challenges, risks and threats,

DOI: 10.4324/9781003298052-15

which can be caused not only by relations among countries, but also by a variety of types of conflicts within states. It is obvious that the globalization process will in a substantial way modify the challenges, risks and threats known today and influence the strategy of how to solve them in the future.

(Slovakian NC 2001a:2)

Regional conflicts, especially in unstable areas such as the Caucasus, uncontrolled migration, international organized crime, terrorism, the activities of foreign special services and the degradation of the environment were all mentioned as security risks. 'The excessive dependence' of Slovakia 'on unstable sources for some basic raw materials and energy and their transport gives rise to a risk, which can grow into a threat not only to economic prosperity and stability, but also to the security of the state', the Council warned (Slovakian NC 2001a:5). The international political-military position of Slovakia was considered favourable. The probability of a large-scale armed conflict had, according the Council, decreased substantially since the end of the Cold War. The Council nevertheless warned that new conflicts still could escalate into armed aggression. In addition, non-military risks could, the Council argued, lead to the necessity of using armed forces to manage the situation. Natural disasters, industrial catastrophes, the proliferation of WMD and the shortage of basic foodstuffs were among the threats considered as having the potential 'trigger a crisis situation which could considerably endanger state security' (Slovakian NC 2001b:4). The Slovakian General Staff (GS) concluded:

> After the likelihood of external military threats decreased due to the fading threat of global confrontation, the Slovak Republic and other countries now face the problem of growing probability of regional conflicts in regions which are in the area of national interest of the Slovak Republic.
>
> (Slovakian GS 2003)

In addition to regional conflicts, the growing potential for the misuse of cybernetic space, radical ideologies, demographic developments including migration, ecological changes and increased struggle for vital resources worried the government. These and other factors were considered as having the potential to cause 'instability and uncertainty, which are seconded by high levels of insecurity, unpredictability, and an increasing potential for the development of unexpected crisis situations', the government argued (Slovakian NC 2005b:3). The growing number of failed states, the negative influence of certain non-state actors, the increased number of intra-state conflicts, the 'increased dependence on vital natural resources and their scarcity, and environmental degradation' all contributed to the 'growing instability, uncertainty, and unpredictability in the world', the government argued (Slovakian NC 2005a:4). In 2013, the government observed that space and 'its use by sophisticated but violable infrastructure represents an area of international co-operation, but also rivalry'. Higher 'utilization of space for military purposes is likely', the government concluded. The likelihood of an extensive

152 *The empirical exploration*

conventional armed conflict in the Euro-Atlantic area was, in the near future, considered to be low, 'but it cannot be completely excluded', the government warned (Slovakian MoD 2013:47–48). The increased challenges and threats in the cyber domain worried the government (Slovakian NC 2015; Slovakian NSA 2015). However, other developments in the strategic environment were perceived as even more demanding. Minister of Defence Peter Gajdoš observed:

> Developments in Ukraine in 2014 presented a fundamental change in the security environment in Europe, which steadily grew worse after the destabilisation of the Greater Middle East and a concomitant growth of security risks for the Slovak Republic and her Allies. The era of a stable security environment with low risks in the external environment, considered as a matter of course by some, is over.
>
> (Slovakian MoD 2016:7)

One year later, the government concluded that the 'increasing instability and tensions between states and the concentration of large and modern conventional military capabilities near the eastern borders of NATO or EU member states can lead to a crisis situation threatening their member states' and the subsequent fulfilment of Slovakia's international obligations (Slovakian NC 2017:3–4). Since 2017, the government continued to observe some significant threats to the security of Slovakia, including the COVID-19 pandemic, the continued erosion of multilateralism, the strengthened competition between powers and hybrid threats (Slovakian Government 2020).

13.3 Ends

Following the expansion of NATO in Central Europe, the Slovakian NC observed that Slovakia 'still has not sufficient external international institutional security guarantees based on a contractual foundation of common defense'. Especially a Slovakian membership in NATO was hence considered essential (Slovakian NC 2001a:2). The NC presented a list of vital and important interests of Slovakia. The latter group included (1) peace and stability in the world, (2) the prevention of tensions and crises and, in case of unsuccessful prevention, their early and effective resolution, (3) good relations with immediate neighbours, (4) the development of all forms of mutually advantageous regional cooperation, (5) internal political stability, (6) the achievement of a dynamic national transition to a market economy, (7) social peace and stability in the Slovakian society, and (8) environmental security. The former group consisted of some overarching ends:

- guaranteeing the sovereignty, territorial integrity and cultural identity of the Slovak Republic;
- preserving and developing the democratic foundations of the Slovakian state as well as its internal security and order;
- protecting the lives and safeguarding the health of Slovakia's citizens;

- ensuring and developing the economic, social, environmental and cultural aspects of the Slovakian society;
- protecting important state infrastructures;
- reserving peace and stability in Central Europe; and
- achieving full membership of Slovakia in both NATO and the EU.

(Slovakian NC 2001a)

The Slovakian citizen's impression of the state's credibility fulfilling Slovakia's obligations arising out of international treaties and agreements, as well as the state's capability to resolve critical situations, including facing threats of violence and aggression, was mentioned as a foundation for the security and defence policy of Slovakia (Slovakian GS 2003). After entering NATO and the EU, the Slovakian government announced that the security interests of Slovakia now also included 'strengthening strategic transatlantic partnership, co-guaranteeing the security of its allies', as well as improving the effectiveness of the international organisations 'which the Slovak Republic is a member of, and supporting NATO and the EU enlargement' (Slovakian NC 2005a:2). The basic goals of the Slovakian defence policy providing security for its citizens and guaranteeing defence of the state was declared to

> derive from the needs of its citizens and of the state. They are based on the values of freedom, peace, rule of law, democracy, prosperity, and observation of basic human rights. These security interests of the Slovak Republic can be characterized as stable in the longterm although pursued in a dynamically changing security environment.
>
> (Slovakian NC 2005b:3)

In 2013, the basic and long-term objective of the Slovakian defence policy was declared to be ensuring the security of Slovakia's citizens and national defence 'based on the capacities of the state, the means of collective defence within NATO and utilizing instruments of the Common Security and Defence Policy of the EU' (Slovakian MoD 2013:57). The government admitted that the Slovakian economy was largely dependent on the import of strategic raw materials, especially energy, from one external supplier, and pledged to increase its measures to diversify import in this regard. Ultimately, the objective was to decrease the risks of being exploited by means of 'pressure from the supplier and the possible threat of stopping the supply of raw materials' (Slovakian Government 2018:3).

13.4 Means

In order to meet the political-military and other criteria for achieving a membership in NATO, the government observed the necessity of undertaking a fundamental reform of the Slovakian armed forces 'to enhance their readiness and their capability for State defense, as well as to increase their interoperability with the Armed Forces of the member states' of NATO (Slovakian NC 2001b:3). Active

154 *The empirical exploration*

participation in international military cooperation, including contributing to military operations under the auspices of international organizations, was hence considered a precondition. At the same time, Slovakia was to secure an effective defence of the Republic by its own forces. Maintaining a high readiness force, 'consisting of at least a battalion-sized element, with appropriate air elements and other supporting units', was hence given priority. The forces at lower readiness were to consist of at least a brigade-sized element. In addition, maintaining appropriate resources in order to provide for the continuous defence of Slovakian airspace as well as 'the capability to receive allied reinforcements [. . .] and to provide other Host Nation Support activities' were the focus of the transformation (Slovakian NC 2001c:6). Manoeuvre warfare and the ability to conduct joint operations as well as to participate in multinational operations were central in the GS's approach to the transformation of the armed forces (Slovakian GS 2003).

The government admitted the changes to be undertaken within the Slovakian armed forces were influenced not only by the ability to defend Slovakia and to contribute to NATO collective defence and EU military capabilities. Other influences were taken from the necessity of having a preparedness for participation in international conflict prevention and crisis resolution in unstable regions of the world. Consequently, deployment and sustainability of the armed forces 'within a wide range of operations led by NATO and other international organizations or coalitions' were considered essential. Notably, the 'required usability level' for the latter wide spectrum capacity was to be attained only after 2015. Until 2015, the ambition was limited to 'adequately contribute to conflict prevention and crisis situation management'. For high-intensity allied operations under NATO collective defence, the ambition was to contribute with land forces up to the size of a brigade group. For geographically unlimited NATO-led multinational joint operations, the goal was to provide land forces up to the size of a battalion group without rotation. Regarding NATO or EU-led multinational peace support operations, the ambition was to provide and sustain 'land forces or its cost equivalent up to the size of a mechanized battalion'. When it comes to 'peace-support and humanitarian operations under the leadership of the UN, EU, or international coalitions', the goal was to provide 'land forces or its cost equivalent of long-term sustainability up to the size of a company' (Slovakian NC 2005b:6–7). Regarding the development of the air force, the short-term priority was 'adequate participation in defence of the NATO integrated air space'. In the long-term perspective, the air force was to 'modernize its aviation assets, ground based air defence systems, and surveillance means' in order to 'allow admittance into NATO anti-ballistic defence' as well as achieving the ability to provide fire support to land forces. Moreover, the air force was to increase its 'transport capacities necessary for Land Force support in operations outside the state territory' (Slovakian NC 2005b:10).

'We fully realize that our ability to defend ourselves is closely linked to the economic security of the state. We therefore proposed, while taking the need for rigorous fiscal consolidation into account, to implement the development strategy in two stages', Minister of Defence Martin Glváč announced in 2013. In the

Table 13.1 Main military resources of Slovakia.

	2000	2010	2020
Army brigade	2	2	2
MBT	275	245	30
ACV/APC/IFV	518	505	350
Combat aircraft	84	22	23

Source: International Institute for Strategic Studies (2000, 2010, 2020).

first stage, the ambition was limited to the equipping of one battalion of the land forces. 'In the second stage, in the period of 2016–2024, the main output will be the rearmament of a mechanized brigade', Minister Glváč continued (Slovakian MoD 2013:9). Looking back at the first two decades of independence and the gradual rebuilding of the Slovakian armed forces following the break-up of Czechoslovakia, Slovakia 'did not manage to meet all long-term transformational goals. Perhaps the most pressing problems are associated with the slow modernization of armament', General Peter Vojtek, Chief of the Slovakian GS, declared (Slovakian MoD 2013:13). The government explicitly admitted that the transformation of the armed forces had failed in three key areas. Regarding military personnel, the reduction of numbers by one-third, partly through the introduction of professional service, had not prevented the aging of military personnel and understaffing of units. With regard to equipment, 'no armaments project concerning major military equipment was implemented, resulting in more than 70% of ground equipment being past its life cycle; in recent years modernization has basically stopped'. Finally, regarding military capabilities, the level of interoperability reached only 54 percent on average according to NATO standards. 'This puts in question the ability of the Slovak Republic to defend itself as well as the quality of fulfilment of international commitments, especially in terms of preparation for future conflicts', the government concluded. One of the underlying causes identified was 'a long-term imbalance and a growing gap between the tasks of national defence and resources devoted to it'. In addition, due to the economic crisis, the defence expenditures had gradually been reduced to 1.1 percent of GDP in 2012 (Slovakian MoD 2013:17–18).

The government announced that the ambition from 2001, that is the armed forces having a personnel strength of 62,000, had to be adjusted to 32,000 organised in six brigade equivalents in the state of war. The ambitions were to have one mechanized battalion group under NATO standards as well as eliminating the lack of tactical air transport capability by 2016. Moreover, the goal was having one mechanized brigade as well as multipurpose tactical aircraft, medium-sized transport aircraft and combat helicopters fully operational by 2024 (Slovakian MoD 2013). Three years later, the government announced its ambition to achieve the level of 1.6 percent of GDP on defence expenditures in 2020. The ambition now was to have the mechanized brigade operational already by 2018. Other prioritised areas included HNS and special forces (Slovakian MoD 2016).

156 *The empirical exploration*

13.5 Ways

Active participation in international peace efforts was considered as enabling Slovakia 'to contribute to the shaping of the world security environment' by the Slovakian NC in 2001. The Council recognized the need to increase Slovakia's contribution 'towards preventing conflicts from emerging and assisting in their resolution should they emerge' (Slovakian NC 2001a:4). In addition to NATO and the EU, the OSCE and the cooperation within the framework of the four Visegrád-countries were hence considered of special importance. Slovakia considered NATO, the Council clarified, 'to be the only effective organization capable of guaranteeing security and peace in Europe' (Slovakian NC 2001b:2). Consequently, the Slovakian approach was based on two fundamentals. First, regarding the national defence, by ensuring 'a credible deterrence and, if this fails, defend the sovereignty, independence and territorial integrity of Slovakia, and that of its allies, against any external threat of armed aggression'. Hence, the Slovakian armed forces were to be 'prepared to conduct such operations unilaterally or as a coalition member'. Second, by participating in a broad scope of multinational military operations ranging from peace enforcement to humanitarian assistance (Slovakian NC 2001c:5). Two years later, this core essence remained, although it was differently expressed. Regarding the national defence, ensuring a plausible deterrence and, in case of failure, defending Slovakia and its allies against an external armed attack was hence the aim. The operations were to be conducted with a 'maneuverist approach' both independently and as part of a coalition. Slovakian armed forces were to 'be prepared to perform their tasks in multinational operations' under operational command 'by the multinational force commander of the UN, OCSE, NATO or EU', the GS declared (Slovakian GS 2003:37).

Once becoming a member of both NATO and the EU, fulfilling Slovakia's commitments as a member in these organisations was given a more predominant position in the elaborations. 'Taking into consideration the projected development of the security situation, the Armed Forces shall be most likely committed to peace-support and anti-terrorism operations, primarily focusing on crisis prevention and stabilization efforts', the government concluded. Until the end of the year of 2010, the ambition was 'to prepare its Armed Forces for participation at least in two simultaneous operations. The priority shall be to attain readiness, firstly for NATO-led operations and secondly for peace-support operations led by international organizations' (Slovakian NC 2005a, 2005b:7–8). In 2013, the government reviewed Slovakia's contributions to international peace efforts since the establishment of the independent Slovak Republic in 1993. Slovakian 'armed forces participated in more than thirty operations and missions abroad led by NATO, the EU, the UN or within the framework of international coalitions in different countries on three continents', the government summarised. On average, slightly fewer than 600 soldiers had been sustainably deployed during recent years. In view of the limited national defence capabilities, Slovakia was considered as being able to respond effectively to most 'security threats and challenges only through playing an active role within international organizations' (Slovakian MoD 2013:32

and 51). Throughout the explored period, the government stressed the special relationship with Czechia and the close partnerships with Austria, Hungary and Poland. In addition, the Visegrád cooperation was given specific attention (see, for example, Slovakia Ministry of Foreign and European Affairs (MFEA) 2020).

13.6 Conclusions: Slovakian strategy

We conclude that the Slovakian alignment strategy corresponds with our criteria for the *multiple-courting* hedging strategy. Throughout the twenty-first century, not only NATO but also the EU and the UN have been in focus regarding international military operations. Arguably, the former has been prioritised. In addition, the Slovakian government has established several bi-, tri- and other multilateral arrangements. Hence, especially the other three Visegrád states and Austria have received most of the attention. Regarding the ends of the military strategy, the government presented different categories of interests. Since most of the ends of the prioritised category, the vital interests, clearly have to do with not only sovereignty and territorial integrity but also cultural identity, we conclude that *survival* is at the core of the Slovakian strategy. Active participation in international military cooperation as well as relevant contributions to multilateral military operations was initially considered a precondition gaining entrance to NATO and to the EU. Consequently, the government launched a transformation aiming at increasing the capacity for expeditionary warfare.

A decade later, the government officially declared the transformation a failure. Despite the economic difficulties and the Russian war against Ukraine, we argue that the international focus, once again, was prioritised when the government launched a second attempt of transforming the military. Although we observe the ambitions in the end of the explored period, which included increasing defence expenditures and speeding up the tempo for having the mechanized brigade operational, we argue that *expeditionary warfare* has been the focus over time. The other leg of the Slovakian concept, both nationally and collectively, focused on deterrence. Clearly, a *multilateral approach* regarding the ways has been the preferred option. Active participation in international peace efforts has been the guiding star throughout the period explored. Even if NATO is considered to be the most effective organisation in this regard, the EU, the OSCE and the Visegrád cooperation have also been emphasised as being of special importance. Bilaterally, the relationship with Czechia has a unique position in the Slovakian strategy, but the partnerships with Austria, Poland and Hungary have also been prioritised.

Table 13.2 Slovakian strategy.

Alignment strategy	Military strategy		
	Ends	Means	Ways
Multiple-courting	Survival	Expeditionary warfare	Multilateral approach

Obviously, we agree with Pavol Kanis in arguing that Slovakia wanted not only to receive security through NATO and the EU but also to 'provide its contributions' internationally (Kanis 2000:32). Zdeněk Kříž and Martin Chovančík provide additional support to our findings regarding the goal of the military reforms, that is 'to establish military forces capable of expeditionary operations' (Kříž and Chovančík 2013:61). Together with Jana Urbanovská, Kříž provides additional support regarding our conclusion on the Slovakian alignment strategy (Kříž and Urbanovská 2013; see also Fischer 2019). We also agree with Matúš Korba that the lack of budget resources created a problem when implementing the transformation of the armed forces (Korba 2001). Arguably, the detailed findings regarding the status of the different units of the armed forces, presented by Michał Fiszer, supports our conclusions. We especially appreciate his description of the aircraft of the Slovakian air force and the capacity to perform recognisance rather than attack missions (Fiszer 2005).

We also appreciate the arguments provided by Ivo Samson, that focuses not only on Slovakia's military cooperation with the Visegrád countries but also on its economic dependency on Germany (Samson 2005). Together with Jozef Ulian, Samson criticises the Slovakian authorities for missing years of opportunities to reform not only the military but also the whole security sector. These findings are in line with our own (Samson and Ulian 2011). Potentially, Juraj Marušiak provides interesting explanations why the Russian war against Ukraine did not fully result in a reorientation towards national defence in Slovakia. Nevertheless, his arguments support our findings in this regard (Marušiak 2015). Elemír Nečej and Samuel Žilinčík provide additional insights about the Slovakian perceptions of Russia, hence also supporting our conclusions (Nečej and Žilinčík 2017; see also Maksymets 2018). Notably, Milan Sopóci and Marek Walancik argue differently when it comes to the consequences of the Russian aggression (Sopóci and Walancik 2015, 2016).

Notes

1 We would like to express our gratitude to Lieutenant Colonel Jörgen Marqardsen, Embassy of Sweden, Bratislava, for his support.
2 Mikuláš Dzurinda of the Democratic and Christian Union–Democratic Party (SDKÚ-DS) served as prime minister 30 October 1998–4 July 2006. Robert Fico of the Direction–Social Democracy (SMER-SD) served as prime minister 4 July 2006–8 July 2010 and again 4 April 2012–22 March 2018. Iveta Radičová of the SDKÚ-DS served as prime minister 8 July 2010–4 April 2012. Peter Pellegrini of the SMER-SD served as prime minister 22 March 2018–21 March 2020. Igor Matovič of the Ordinary People and Independent Personalities (OL'aNO) served as prime minister 21 March 2020–1 April 2021. As of 1 January 2022, Eduard Heger of the OL'aNO has served as prime minister since 1 April 2021. See www.vlada.gov.sk//government-of-the-slovak-republic/.

Bibliography

Fischer, Dušan (2019). 'Relations between the United States and Slovakia: Friends and Allies between 1989 and 2017' in Anna Péczeli (ed). *The Relations of Central European Countries with the United States*. Budapest: Dialóg Campus.

Fiszer, Michał (2005). 'Slovakia Works toward Integration with NATO' *Journal of Electronic Defense* Volume 28, Issue 1.
Gorys, Erhard (1996). *Czech and Slovak Republics*. London: Pallas Athene.
International Institute for Strategic Studies (IISS) (2000). *The Military Balance 2000–2001*. Oxford: Oxford University Press.
——— (2010). *The Military Balance 2010*. London: Routledge.
——— (2020). *The Military Balance 2020*. London: Routledge.
Kanis, Pavol (2000). 'The Path Ahead for the Slovak Armed Forces' *Military Technology* Volume 24, Issue 4.
Kirschbaum, Stanislav (1995). *A History of Slovakia: The Struggle for Survival*. Basingstoke: Macmillan.
Korba, Matúš (2001). 'Civil-Military Relations in Slovakia from the Perspective of NATO Integration' *Slovak Foreign Policy Affairs* Volume 2, Issue 2.
Kříž, Zdeněk and Martin Chovančík (2013). 'Czech and Slovak Defense Policies since 1999: The Impact of Europeanization' *Problems of Post-Communism* Volume 60, Issue 3.
Kříž, Zdeněk and Jana Urbanovská (2013). 'Slovakia in UN Peacekeeping Operations: Trapped between the Logic of Consequences and Appropriateness' *Journal of Slavic Military Studies* Volume 26, Issue 3.
Liddell-Hart, Basil (1997). *History of the Second World War*. London: Papermac.
Maksymets, Vira (2018). 'Security Policy of the Slovak Republic in the Context of Intensifying the Hybrid Actions of the Russian Federation' *Grani* Volume 21, Issue 3.
Marušiak, Juraj (2015). 'Russia and the Visegrad Group: More Than a Foreign Policy Issue' *International Issues & Slovak Foreign Policy Affairs* Volume 24, Issue 1–2.
Nečej, Elemír and Samuel Žilinčík (2017). *Analysis of the Draft of Security Strategy of Slovak Republic 2017*. Bratislava: Strategic policy institute.
Samson, Ivo (2005). 'Slovakia' in Tom Lansford and Blagovest Tashev (eds). *Old Europe, New Europe and the US: Renegotiating Transatlantic Security in the Post 9/11 Era*. Abingdon: Routledge.
Samson, Ivo and Jozef Ulian (2011). 'Problems of Security Sector Reform in Slovakia' *International Issues & Slovak Foreign Policy Affairs* Volume 20, Issue 3.
Slovakia Ministry of Foreign and European Affairs (MFEA) (2020) *Foreign and European Policy*.
Slovakian General Staff (GS) (2003). *Doctrine of the Armed Forces of the Slovak Republic*.
Slovakian Government (2018). *Concept for the Fight of the Slovak Republic against Hybrid Threats*.
——— (2020). *Proposal for the Preparation of the Security Strategy of the Slovak Republic*.
Slovakian Ministry of Defence (MoD) (2013). *White Paper on Defence*.
——— (2016). *White Paper on Defence*.
Slovakian National Council (NC) (2001a). *Security Strategy of the Slovak Republic*.
——— (2001b). *Defence Strategy of the Slovak Republic*.
——— (2001c). *Military Strategy of the Slovak Republic*.
——— (2005a). *Security Strategy of the Slovak Republic*.
——— (2005b). *Defence Strategy of the Slovak Republic*.
——— (2015). *Cyber Security Concept of the Slovak Republic 2015–2020*.
——— (2017). *Defence Strategy of the Slovak Republic*.
Slovakian National Security Authority (NSA) (2015). *Action Plan for the Implementation of the Cyber Security Concept of the Slovak Republic for 2015–2020*.

Sopóci, Milan and Marek Walancik (2015). 'Security and Defence Sources for the Armed Forces of the Slovak Republic' *Journal of Defense Resources Management* Volume 6, Issue 1.
——— (2016). 'Positive Trends in Defense Resources for the Armed Forces of the Slovak Republic' *Journal of Defense Resources Management* Volume 7, Issue 1.
Wallace, William (1977). *Czechoslovakia*. London: Ernest Benn Ltd.

14 The strategy of Slovenia[1]

The primary sources analysed in this chapter are the strategic defence reviews of 2004 and 2016, the defence sector strategic review of 2009, the annual reports on defence, the DWP of 2020, the NDS of 2012 and the NSSs of 2010 and 2019. In addition, the strategies for Slovenia's participation in international operations and missions and for cyber security have been studied.[2]

14.1 Historical background

Throughout history, the territory of present Slovenia has been part of many different states, including the Byzantine Empire, the Holy Roman Empire, the Kingdom of Hungary, the Republic of Venice, the French Empire (as part of the Illyrian Provinces) and the Austro-Hungarian Empire. Notably, the term 'Slovenia' was not used before the early nineteenth century. In late 1918, following the end of WWI, the Slovenes exercised independence for the first time when co-founding the State of Slovenes, Croats and Serbs, in 1929 renamed the Kingdom of Yugoslavia. In April 1941, the Axis powers invaded Yugoslavia leading to Germany, Hungary and Italy portioning and occupying Slovenia. After WWII, Yugoslavia regained its independence, and the monarchy was replaced with a federal republic. In June 1991, Slovenia declared independence from Yugoslavia. The Slovenian War of Independence was brief and took only ten days without the extreme violence that came to characterise most of the other parts of the former federation (see, for example, Liddell-Hart 1997: Gow and Carmichael 2000; Ferfila *et al*. 2004).

14.2 Strategic environment

In 2004, the Slovenian government observed 'the increasingly diverse, unpredictable and constantly changing international environment, the ever more frequent and intensive regional and local conflicts, and the changing nature and priorities of security risks and threats' (Slovenian MoD 2004:7). 'The importance of traditional military threats is declining, although they cannot be entirely ruled out', the government argued (Slovenian MoD 2004:8). Terrorism, drug trafficking, organised crime, mass migration, and proliferation of WMD were considered to be

DOI: 10.4324/9781003298052-16

particularly notable non-military sources of threat. Potential points of conflict in Slovenia's immediate region were considered to be the Balkans, the Middle East and the Caucasus. Slovenia must 'actively endeavour to ensure stability in the region and beyond, because stability in the regional and global economic environment has a positive impact on the country's economy', the government explained (Slovenian MoD 2004:10). 'Armed conflict within the wider region may occur, but in a limited form. The likelihood of military conflict between the major powers having an impact on Slovenia is minimal', the government concluded (Slovenian MoD 2004:27).

In 2009, the elaborations often took NATO rather than Slovenia as the point of departure. 'New security challenges and threats that pose a direct and indirect threat to [the] security of NATO member countries are usually non-military by nature but may also involve military implications', the government observed. 'Large-scale conventional aggression against the Alliance remains unlikely in the future. However, possible future attacks with military and non-conventional means can also affect the Euro-Atlantic region', the government concluded (Slovenian MoD 2009a:7). In the NSS issued in 2010, the government observed that the

> modern international security environment is complex, interdependent, subject to unpredictable changes and of global proportions. Due to the creation of new global centres of power and the re-emergence of old ones, its multipolar nature is now being strengthened. All this reflects in the security threats and risks.
>
> (Slovenian Government 2010a:8)

The government also elaborated on cyber threats, the activities of foreign intelligence services, climate change, scarcity of strategic resources and the new forms and nature of conflicts. The possibility of armed conflicts between states in the Euro-Atlantic area had, the government concluded, diminished significantly. In the short- and medium-term, 'Slovenia is not directly exposed to military threats; however, a significantly altered international and regional political and security environment may result in exposure to such threats', the government stressed (Slovenian Government 2010a:20). The government also observed the increased asymmetrical and multilayered character of the modern threats and challenges. In addition to the regions previously prioritised, the government also stressed the importance of Eastern Europe and the Caucasus region (Slovenian Government 2010b).

In 2011, the government concluded that even if the likelihood of interstate war in the Euro-Atlantic region remained very low, military threats nevertheless could arise as spillovers from local and regional instabilities elsewhere. 'Moreover, contemporary threats are increasingly becoming hybrid in their form,' the government concluded. Besides individual countries, non-state and transnational actors were considered to be the likely source of future threats to the international security environment. The government also observed the increased importance of both

The strategy of Slovenia 163

space and cyberspace (Slovenian MoD 2011a:6). In 2016, the increasingly rapid development of information and communication technologies worried the government. The trend in the use of these technologies 'for political, economic and military supremacy is becoming more and more pronounced. Undoubtedly, cyberattacks are among the most significant security threats to the modern world, and therefore, cyber security has become an important, integral part of national security', the government argued (Slovenian Government 2016:3). 'After decades of easing of tensions in the international security environment, these are increasing once more', the government observed in 2019. Not least the increased instability in Slovenia's immediate neighbourhood was considered worrisome (Slovenian Government 2019:8).

14.3 Ends

In 2004, the government declared the main strategic goals of its national security policy to be 'to ensure national integrity and prosperity by means of an active role in the international community and in the globalised world' and the 'protection and preservation of the national identity' (Slovenian MoD 2004:17). Protection of human rights and natural resources, respect for international law, as well as ensuring international economic and social development were the declared aims of the government. Strengthening national security, reducing exposure to risks, ensuring the security and welfare of the citizens, and preventing and managing crises and conflicts were also central ends. Other core objectives were to consolidate not only peace, security and stability but also Slovenia's position and reputation in the international community. 'A definitive settling and stabilising of the political and security situation in the former Yugoslavia and the wider region is of vital importance for Slovenia', the government announced (Slovenian MoD 2004:14). 'With the Republic of Slovenia joining NATO and the EU, the key objectives of defence policy from the turn of the century were achieved', the government admitted in 2009 (Slovenian MoD 2009a:6).

In the NSS presented in 2010, the government distinguished the national interests as vital or strategic. The former category included the preservation of not only the independence, sovereignty and territorial integrity of the Slovenian state but also of the identity, culture and autonomy of the Slovenian nation. The latter category included international recognition of and respect for Slovenia's territorial borders, the efficient functioning of Slovenia's democratic political system and security for Slovenia's people. Respect for human rights and fundamental freedoms as well as the strengthening of the rule of law and of the welfare of both the state and the people, were also considered as strategic interests. The protection of the rights and development of the indigenous Slovenian minorities in neighbouring countries was another strategic interest expressed by the government (Slovenian Government 2010a).

In 2011, the government made clear that Slovenia 'will defend its independence, sovereignty and territorial integrity with all available means and methods that are in compliance with the provision of the international and humanitarian

164 *The empirical exploration*

law' (Slovenian MoD 2011a:16). 'Maintaining independence, sovereignty and territorial integrity as well as inviolability of Slovenia's internationally recognized borders and national territory' was declared the core objective in the defence strategy of 2012 (Slovenian MoD 2012b:4). The protection of the rights and prosperity of Slovenian indigenous ethnic communities in neighbouring countries was expressed as a vital national interest also in the NSS issued in 2019. So were 'the protection and strengthening of constitutional principles, the national identity, culture, and authenticity of the Slovenian nation' (Slovenian Government 2019:6). The government also considered Slovenia as being an integral part of the Euro-Atlantic political, economic, security and cultural environment as a core national interest (Slovenian MoD 2020).

14.4 Means

As of 2003, the wartime strength of the Slovenian armed forces was about 18,000 personnel (Slovenian MoD 2004). Regarding participation to NATO-led operations, the government announced that Slovenia would contribute with

> a motorised infantry company with rotation capability for cooperation in all NATO tasks, with limitations relating to air defence and communications. By the end of 2006 we will be capable of providing a motorised infantry battalion. By the end of 2009 we will be able to provide a fully deployable and supported battalion combat group, and by the end of 2012 a fully deployable and supported battalion combat group with rotation capability [. . .]. In the near future, Slovenia's contribution to the joint military capabilities of the EU will remain the same as its contribution to NATO.
> (Slovenian MoD 2004:19–20)

In addition, the government pledged to make available for NATO or the EU an NBC battalion and other minor capabilities. The government presented its ambition to participate, within the framework of the Quadrilateral, that is Croatia, Hungary, Italy and Slovenia, in NATO's MLF with a contingent in a brigade of the high-readiness forces. Notably, the Slovenian armed forces 'will continue to be organised as a unified army not divided into services', the government declared. The structure of the armed forces 'will provide for the maintenance of the out-of-area tactical group in rotation, for which a minimum of three battalions is required', the government concluded. Moreover, an airbase 'with suitable host nation support for Alliance requirements' including the capacity for air traffic control over Slovenian territory was to be provided. In addition, Slovenian armed forces were to 'ensure safe access and protection of the maritime waterways and the port of Koper' (Slovenian MoD 2004:35–36). The goal of the transformation was, by 2015, to have developed three motorised battalions, one light mountain battalion, an artillery battalion, an engineer battalion, an air defence battalion, an NBC battalion, a military police battalion, a communications battalion, an airspace control battalion, a reconnaissance battalion and a helicopter battalion. The armed forces 'will have up to 14,000 personnel, of which a minimum of 8,500 will be professionals', the government declared (Slovenian MoD 2004:40).

In 2008, the first high-readiness battalion battle group as well as the air defence and aviation brigade command became operational. The integration of Slovenian armed forces into NATO's command structure had led to the affiliation of a battalion battle group to the NATO Rapid Deployable Corps based in Italy (NRDC-It) and the NBC battalion into MNC-NE with its headquarters in Poland. The government admitted that the 'armoured vehicle project did not progress as planned' (Slovenian MoD 2009b:42). The personnel strength of the armed forces was adjusted to 10,000 service members in the planned structure. The main units of the armed forces were the three brigades. The bulk of the 1st Brigade was constituted by three motorised battalions. The 72nd Brigade consisted, amongst other units, of a mountain battalion and an armoured battalion. The Air Defence & Aviation brigade organised the airbase, the helicopter battalion and the air defence battalion. In addition to the independent reconnaissance and military police battalions, a special forces unit was organised. Regarding the latter brigade, the government clarified that the 'procurement of transport aircraft has not been realized' (Slovenian MoD 2009a:17). During 2009, both the light battalion battle group and the NBC battalion had been NATO certified, Minister of Defence Ljubica Jelušič announced (Slovenian MoD 2010). The government considered participation in international operations as an important tool for acquiring the knowledge and skills necessary for developing the national military capabilities (Slovenian Government 2010b).

In 2011, the government admitted that the supply of key equipment to the motorised battalions did not allow the formation of the planned medium-weight battalion battle group. Notably, the implementation of the plan was expected to be delayed for four years. Consequently, only a light battle group could be organised. On the other hand, the delivery of a multipurpose patrol boat was considered to strengthen Slovenia's defence capability at sea (Slovenian MoD 2011b). The government admitted that 'in addition to the existing motorised infantry battalion group', the 'medium infantry battalion group' was not to be developed until 2020. The armoured capabilities were hence to consist of an augmented tank company (Slovenian MoD 2011a:18). The government also announced that Slovenia 'will not establish its own aircraft capabilities for air policing in the Slovenian airspace, but will seek appropriate solutions within NATO'. As a result, the Slovenian training aircraft were to 'be removed from operational use'. Tactical movement with up to one infantry company in a single airlift would, however, be accomplished with helicopters (Slovenian MoD 2011a:36). In the context of national defence, the ambition was to 'generate task-force tactical units up to brigade level', while in Article 5 operations outside Slovenia, 'capabilities for the formation of a medium infantry battalion battle group will be generated in addition to light infantry battalion battle group capabilities', the government announced in 2012 (Slovenian MoD 2012b:54). However, the 'financial crisis and the consequent reduction in investments resulted in significantly reduced equipping' of the armed forces, Minister of Defence Aleš Hojs admitted (Slovenian MoD 2012a:5). 'The financial

166 *The empirical exploration*

Table 14.1 Main military resources of Slovenia.

	2000	2010	2020
Army brigade	8	1	2
MBT	46	70	14
ACV/APC/IFV	122	124	115
Combat aircraft	0	9	9
Principle surface combatant	0	1 patrol craft	2 patrol craft

Source: International Institute for Strategic Studies (2000, 2010, 2020).

resources available in 2012 did not allow for the development and building of the [armed forces'] capabilities, provided for in the medium- and long-term planning documents', which his successor, Roman Jakič, had to admit the following year as well (Slovenian MoD 2013:5; see also Slovenian MoD 2014:13).

Despite the economic difficulties, the government announced its ambition developing the medium battalion battle group into a mechanized battalion battle group (Slovenian MoD 2016a). 'The main development goal in the medium term 2018–2023 is the formation of two medium battalion battle groups', the government declared. The government also clarified that the increase of 'the defence budget to approximately 2% of GDP' was to take place 'over the long term until 2027' (Slovenian MoD 2016b:6 and 8). By 2016, the key equipment of the armed forces consisted of 18 105 mm howitzers, a Roland air defence missile system, 39 BVP M80A IFVs, ten M-84 MBTs and 30 T-55S MBTs (Slovenian MoD 2017). In 2018, the government still strived for the formation of two medium battalion battle groups. Notably, the enlargement and modernization of the Cerklje ob Krki airfield, which had taken place over several years, had to be financed from NATO funds (Slovenian MoD 2018). In 2018, the strength of the armed forces consisted of slightly less than 7,500 personnel with 6,628 in the active component and 828 in the contract reserve component (Slovenian MoD 2019).

14.5 Ways

As of 2004, approximately 200 troops altogether participated in the NATO-led operations SFOR, KFOR and ISAF. 'Slovenian defence policy is oriented towards an active and constructive role, particularly within the various bodies of the United Nations, NATO and the European Union', the government declared. The OSCE was also mentioned in this regard (Slovenian MoD 2004:8). The government listed a number of missions that the armed forces were to be able to carry out. The list included national military defence operations, collective defence operations and crisis response operations. The latter mission included the whole spectrum from preventive diplomacy through peacekeeping as well as peace enforcement to post-conflict peace building. Evacuation of Slovenian citizens abroad in peacetime as well as in crises and wartime was an additional mission. 'The missions and tasks of the [armed forces] determine their size and capabilities', the government argued (Slovenian MoD 2004:35).

In 2008, Slovenia had a reduced battalion deployed to KFOR. In addition, Slovenian armed forces were deployed to ISAF within the frames of the Quadrilateral unit to MLF. Moreover, personnel were deployed to the EU-led operations in Bosnia and Hercegovina and to Chad and the Central African Republic as well as, with a squad-level unit, to UN Interim Force in Lebanon (UNIFIL) in Lebanon. Altogether, about 1,300 military personnel were deployed abroad (Slovenian MoD 2009b). In addition to NATO and the EU, the neighbouring countries, that is Austria, Italy and Hungary, were considered as Slovenia's most important strategic partners not least through the provision of the joint military capability within the frames of MLF. 'Slovenia's bilateral cooperation with all neighbouring and some key European countries [. . .] is well-developed', the government announced Slovenian MoD 2009a:19). Regarding the main missions of the armed forces, the government declared that these 'will continue to be the deterrence of military aggression on the Republic of Slovenia in cooperation with allies', as well as the 'contribution to international peace and stability within and beyond Alliance boundaries' (Slovenian MoD 2009a:19). The government gave priority not only to participation in military operations led by NATO and the EU but also to some spatial areas. South East Europe, the Middle East, Central Asia and North Africa were hence considered prioritised areas (Slovenian Government 2010a).

In 2011, the government admitted that, due to the limited national defence resources, the defence of Slovenia would primarily be 'ensured though collective defence and collective security'. The government added that a 'reasonable amount of independence and autonomy in the defence and military areas' nevertheless would be to provide 'an appropriate level of own defence capabilities and readiness'. At the same time, the government pledged to contribute 'various units to NATO response forces and EU battle groups' (Slovenian MoD 2011a:16–18). During 2014, Slovenia reduced its contributions to ISAF from 34 to two military personnel. On the other hand, Slovenia participated with 39 personnel in the Italian-led naval operation Mare Nostrum during one and a half months (Slovenian MoD 2015). Once the mechanized battalion battle group was developed, the government clarified that Slovenia would be 'capable of operating in all potential Alliance operations' (Slovenian MoD 2016a:11). The armed forces have 'a limited ability to achieve its mission and the tasks assigned to it both within and outside the national territory to the extent required across the full spectrum of combat operations', the government admitted in 2017 (Slovenian MoD 2017:16).

14.6 Conclusions: Slovenian strategy

On the one hand, we find it reasonable to argue that the Slovenian alignment strategy corresponds to the criteria of a multiple-courting hedging strategy. Over the past two decades, NATO, the EU and the UN have been the focus militarily. So has, to some extent, the OSCE. On the other hand, the widest scope of

168 *The empirical exploration*

participation of the Slovenian armed forces has been in the NATO-led operations KFOR in Kosovo and ISAF in Afghanistan. These two missions are the only ones to which Slovenia has contributed with armed units of at least company size. The contributions to all missions led by other organisations have only been units of platoon and squad level or by individual staffing. Even if the EU has been regarded as fundamental concerning other aspects of security and Slovenia's economic developments, the EU has not been given as central a role militarily as NATO. Since Slovenia, in addition, is heavily dependent on NATO funding for developing its domestic military infrastructure, we argue that *passive bandwagoning*, that is buck-passing, is a more appropriate label.

Regarding the military strategy, we conclude that the ends have been about the survival of both the Slovenian state and nation. However, we also observe ends going beyond pure survival, striving for recognition internationally. Consequently, we argue that the considerations regarding the ends indicate a balance between *survival* and *status*. When it comes to the means, we conclude that the Slovenian government has had a clear focus on *expeditionary warfare*. Organising an all-voluntary force with a personnel strength of some 8,000 troops simply does not match the needs of an armed force focusing on national defence. Clearly, the Slovenian government had had economic difficulties putting deeds behind the words. Despite the expressed ambitions regarding expeditionary capacity, the contributions to international operations are far from impressive. Since the air force lacks aircraft and hence is focusing on tactical airlifts with helicopters, and since the naval capacity mainly is based on a single boat, we find our conclusion reasonable. These facts also provide arguments for the *multilateral approach* regarding the ways. Simply put, Slovenia has no other options. With a population of about 2 million, the dependency of allies is obvious.

Table 14.2 Slovenian strategy.

Alignment strategy	Military strategy		
	Ends	Means	Ways
Passive bandwagoning	Survival/status	Expeditionary warfare	Multilateral approach

We find the conclusions of Igor Kotnik-Dvojmoc and Erik Kopac regarding the expeditionary profile in line with our findings. Obviously, we do not fully agree regarding the hedging approach towards both NATO and the EU (Kotnik-Dvojmoc and Kopac 2002). We also find the analysis of Joseph Derdzinski regarding ends based on survival and status as supporting our position in this regard. His conclusion on serious reliance on others to fulfil Slovenia's security commitments provides additional support (Derdzinski 2003). Arguably, the findings of Rihard Piskar regarding the importance of both collective security and collective defence go in the same direction as our conclusions (Piskar 2003). We do not agree with Ladislav Lipič when arguing that deterring an armed attack has been the focus in the Slovenian strategic elaborations. Since he is presumably referring to the situation before the Slovenian memberships in NATO and the EU, this indicates a shift from a strategy based on national defence (Lipič 2004). We rather agree with Ljubica Jelušič when arguing that peacetime support to civil authorities has a strong position among the tasks given to the Slovenian armed forces. Arguably, this approach has remained to some degree after the transformation towards an all-volunteer force and expeditionary warfare abroad (Jelušič 2005).

We also find Anton Grizold's conclusions on the transformation of the Slovenian armed forces from a territorial defence force to a professional military in line with our own conclusions in this regard. Moreover, his arguments that NATO presumably expects Slovenia to contribute one battalion with combat and logistic support is convincing and in sharp contrast to what Slovenia actually offers, matching our own findings in this regard. Obviously, we agree with his view on NATO as the preferable partner (Grizold 2008). Daniel Sweeney and Joseph Derdzinski also support this position. They add that Slovenia's geopolitical location argued in favour for NATO (Sweeney and Derdzinski 2010). We find Marjan Malešič and his colleagues' conclusions regarding the early suspension of conscription and the professionalization of the Slovenian armed forces convincing. Their observation regarding the transformation towards expeditionary warfare is interesting. So are their findings on internal diversification within the armed forces between the traditionalists' focus on territorial defence and the modernists' ditto on participation in international operations abroad (Malešič *et al.* 2015; see also Furlan 2013).

We find that the elaborations of Ana Bojinović Fenko and Zlatko Šabič regarding Slovenia's place and direction in the world support our conclusion regarding international status as important aspects of the ends (Fenko and Šabič 2017). Clearly, when Mark Kogoj illuminates that almost 85 percent of Slovenia's expenditures for contributing to international military missions between 1997 and 2016 has been directed to NATO-led operations, he also provides support for our conclusion on bandwagoning. 'This clearly shows Slovenia's foreign policy orientation towards NATO', Kogoj claims, and we fully agree (Kogoj 2019:186). However, we do not fully agree with the predictions of Ryan Hendrickson, and Michael Rudy when arguing that Slovenia will not be a 'free-rider' in NATO. Yes, the contributions have been rather small but the usefulness, at least at the military strategic level, is not that obvious as they assumed it would be (Hendrickson and Rudy 2003).

Notes

1 We would like to express our gratitude to Minister Mateja Kavaš and First Counsellor Andreja Thieke, Embassy of Slovenia, Copenhagen, as well as Brigadier General (ret.) Branimir Furlan for their support.
2 Janez Drnovšek of the Liberal Democracy of Slovenia (LDS) served as prime minister 14 May 1992–7 June 2000 and again 30 November 2000–19 December 2002. Andrej Bajuk of the New Slovenia–Christian Democrats (NSi) served as prime minister 7 June–30 November 2000. Anton Rop of the LDS served as prime minister 19 December 2002–3 December 2004. Borut Pahor of the Social Democrats (SD) served as prime

minister 21 November 2008–10 February 2012. Alenka Bratušek of the Positive Slovenia (PS) served as prime minister 20 March 2013–18 September 2014. Miro Cerar of the Modern Centre Party (SMC) served as prime minister 18 September 2014–13 September 2018. Marjan Šarec of the List of Marjan Šarec (LMŠ) served as prime minister 13 September 2018–13 March 2020. Janez Janša of the Slovenian Democratic Party (SDS) served as prime minister 3 December 2004–21 November 2008 and again 10 February 2012–20 March 2013. As of 1 January 2022, he has served as prime minister since 13 March 2020. See www.gov.si/en/.

Bibliography

Derdzinski, Joseph (2003). 'Slovenia's Contemporary Defense Framework: What Implications for Theory?' *Slovene Studies* Volume 1, Issue 2.

Fenko, Ana Bojinović and Zlatko Šabič (2017). 'Slovenia's Foreign Policy Opportunities and Constraints: The Analysis of an Interplay of Foreign Policy Environments' *Croatian International Relations Review* Volume 23, Issue 79.

Ferfila, Bogomil, Anton Grizold, John Loxley and Paul Phillips (2004). *On the Sunny Side of the Alps: Historical, Political, Economic and Strategic Factors in Independent Slovenia*. Ljubljana: University of Ljubljana (Faculty of Social Sciences).

Furlan, Branimir (2013). 'Civilian Control and Military Effectiveness: Slovenian Case' *Armed Forces & Society* Volume 39, Issue 3.

Gow, James and Cathie Carmichael (2000). *Slovenia and the Slovenes*. London: Hurst & Company.

Grizold, Anton (2008). 'Slovenia's Defense Policy in a Euro-Atlantic Reality' *Mediterranean Quarterly* Volume 19, Issue 3.

Hendrickson, Ryan and Michael Rudy (2003). 'Transforming Slovenia's Military' *Journal of Slavic Military Studies* Volume 16, Issue 4.

International Institute for Strategic Studies (IISS) (2000). *The Military Balance 2000–2001*. Oxford: Oxford University Press.

——— (2010). *The Military Balance 2010*. London: Routledge.

——— (2020). *The Military Balance 2020*. London: Routledge.

Jelušič, Ljubica (2005) 'Domestic Military Assistance: The Case of Slovenia' in Timothy Edmunds and Marjan Malesic (eds). *Defence Transformation in Europe: Evolving Military Roles*. Amsterdam: IOS Press.

Kogoj, Mark (2019). 'Relations between the United States and Slovenia: From U.S. Adverseness to Acceptance and Cooperation' in Anna Péczeli (ed). *The Relations of Central European Countries with the United States*. Budapest: Dialóg Campus.

Kotnik-Dvojmoc, Igor and Erik Kopac (2002). 'Professionalisation of the Slovenian Armed Forces' in Anthony Forster, Timothy Edmunds and Andrew Cottey (eds). *The Challenge of Military Reform in Postcommunist Europe*. Basingstoke: Palgrave Macmillan.

Liddell-Hart, Basil (1997). *History of the Second World War*. London: Papermac.

Lipič, Ladislav (2004). 'Deterring Attacks on Slovenia' *NATO's Nations and Partners for Peace*, Issue 4.

Malešič, Marjan, Ljubica Jelušič, Maja Garb, Janja Vuga, Erik Kopac and Jelena Juvan (2015). *Small, But Smart? The Structural and Functional Professionalization of the Slovenian Armed Forces*. Baden-Baden: Nomos Verlag.

Piskar, Rihard (2003). 'Slovenia and National Security' *Journal of Slavic Military Studies* Volume 16, Issue 3.

Slovenian Government (2010a). *National Security Strategy*.

——— (2010b). *Strategy of the Participation in International Operations and Missions.*
——— (2016). *Cyber Security Strategy.*
——— (2019). *National Security Strategy.*
Slovenian Ministry of Defence (MoD) (2004). *Strategic Defence Review.*
——— (2009a) *Defence Sector Strategic Review.*
——— (2009b). *Annual Report on Defence for 2008.*
——— (2010). *Annual Report on Defence for 2009.*
——— (2011a). *General Long-Term Development and Equipping Programme of the Slovenian Armed Forces Up to 2025.*
——— (2011b). *Annual Report on Defence for 2010.*
——— (2012a). *Annual Report on Defence for 2011.*
——— (2012b). *Defence Strategy.*
——— (2013). *Annual Report on Defence for 2012.*
——— (2014). *Annual Report on Defence for 2013.*
——— (2015). *Annual Report on Defence for 2014.*
——— (2016a). *Annual Report on Defence for 2015.*
——— (2016b). *Strategic Defence Review.*
——— (2017). *Annual Report on Defence for 2016.*
——— (2018). *Annual Report on Defence for 2017.*
——— (2019). *Annual Report on Defence for 2018.*
——— (2020). *White Paper on Defence.*
Sweeney, Daniel and Joseph Derdzinski (2010). 'Small States and (In)Security: A Comparison of Ireland and Slovenia' *Connections* Volume 9, Issue 2.

Part III
Explaining the findings

In this third and last part, Part III, the findings and conclusions of the exploration are summarised. In Chapter 15, the results from the empirical exploration are aggregated. Based on the findings, the explanatory power of each of the intervening variables is tested in Chapter 16. Finally, our overarching conclusions are presented in Chapter 17.

15 The aggregated result of the empirical exploration

So, after exploring the strategies of the 11 new European allies, what picture appears? Has the shared characteristic, that is being part of one or several of the four imperial states that ruled Central and Eastern Europe for centuries, shaped a common strategic behaviour? Have the shared military experiences, that is being a theatre of brutal warring during WWII, established a common view on how to deter potential aggressors from attack? Have the shared political experiences, that is being governed by totalitarian communist regimes during the Cold War, fostered a shared strategic culture? Have these common historical experiences been influential enough to produce similarities in the countries' strategic responses to the systematic pressures they all come under once applying for membership in both the EU and NATO? As shown in Table 15.1, the answers to these questions are that the differences among the 11 cases dominate the picture.

Nevertheless, some similarities can be identified. Bulgaria and Croatia seem, for example, to formulate their defence strategies in exactly the same manner. Despite a given priority to survival among the ends, both favour expeditionary warfare when designing the means. The three Baltic States of Estonia, Latvia and Lithuania have also formulated identical defence strategies. In their cases, the focus on survival is complemented with a priority for means designed for national defence. Finally, regarding Romania and Slovakia, the quest for status, as well as designing the means primarily for expeditionary warfare, creates a unique combination in their strategies. All other four countries have formulated individual strategies, making seven the total number of different favoured defence strategies.

In this chapter, we focus on the similarities and differences among the cases regarding the alignment strategy as well as each of the elements of the military strategy. These similarities and differences are explained in the next chapter.

15.1 Alignment strategy

Regarding the alignment strategy, all but Poland and Slovenia applied the hedging strategy of *multiple-courting*. Perhaps unsurprisingly, all these nine new allies stressed the importance of the relationship with the EU and NATO. However, in all these nine cases, the memberships in these organisations seem not to be

DOI: 10.4324/9781003298052-18

Table 15.1 Strategies of the 11 new European allies.

	Alignment strategy	Military strategy		
		Ends	Means	Ways
Bulgaria	Multiple-courting	Survival	Expeditionary warfare	Multilateral approach
Croatia	Multiple-courting	Survival	Expeditionary warfare	Multilateral approach
Czechia	Multiple-courting	Influence/ survival	National defence	Multilateral approach
Estonia	Multiple-courting	Survival	National defence	Multilateral approach
Hungary	Multiple-courting	Survival/status	National defence	Multilateral approach
Latvia	Multiple-courting	Survival	National defence	Multilateral approach
Lithuania	Multiple-courting	Survival	National defence	Multilateral approach
Poland	Offensive bandwagoning	Survival/status	National defence	Multilateral and unilateral approach
Romania	Multiple-courting	Status	Expeditionary warfare	Multilateral approach
Slovakia	Multiple-courting	Status	Expeditionary warfare	Multilateral approach
Slovenia	Passive bandwagoning	Survival/status	Expeditionary warfare	Multilateral approach

enough. Additional relationships were hence something all strived for even if the preferred partner differed. Arguably, two major approaches are established.

The first, the *military-oriented approach*, complements the EU/NATO memberships foremost with bilateral arrangements with the US. Bulgaria, Estonia, Latvia, Lithuania and Romania all favour this alignment strategy. In the case of Bulgaria, we do not exclude that a shift towards a more US-oriented bandwagoning strategy is about to take place. While the three Baltic States also strive for trilateral cooperation among themselves, Romania is more open-minded in this regard and willing to cooperate with others on an *ad hoc* basis.

The second, the *broader security-oriented approach*, complements the EU/NATO memberships foremost with engagement within the UN system, occasionally including other organisations such as the OSCE and the OECD as well. Croatia, the Czech Republic, Hungary and Slovakia all seem to favour this alignment strategy. Notably, the three latter countries also put emphasis on their cooperation within the Visegrád Four.

Clearly, the fourth Visegrád-country, Poland, has given priority to strategic partnership with the US, hence applying an *offensive bandwagoning* strategy. Despite protests from other NATO and EU members, Polish armed forces contributed to the US-led invasion of Iraq without having a mandate from the UNSC. Moreover, Poland has willingly allowed the US to establish facilities for the US ballistic missile defence programme as well as for conventional forces on Polish soil. However, Poland has also strived for multilateral cooperation with the Baltic States and hence especially Lithuania.

Lastly, regarding Slovenia, we argue that the dependency on NATO funding for developing the Slovenian military infrastructure makes *passive bandwagoning* the preferred alignment strategy. Moreover, NATO led the only two military operations – KFOR in Kosovo and ISAF in Afghanistan – to which Slovenia has contributed armed units of at least company size. Notably, the Slovenian contributions to missions led by other organisations have been even less impressive, consisting only of units at squad level or simply of individual staff officers. Consequently, passing the buck to other NATO members seems to be the most appropriate label of the Slovenian alignment strategy.

15.2 The ends element of the military strategy

Notably, all 11 cases but Romania gave, in way or another, priority to *survival*. While most of the cases, that is Bulgaria, Croatia, Estonia, Latvia, Lithuania and Slovakia, solely focused on survival, Czechia, Hungary, Poland and Slovenia put emphasis on survival in addition to one of the other two main ends.

Regarding the countries focusing solely on *survival*, a division among the cases can be identified. The first group, consisting of Bulgaria, Croatia and Slovakia, related survival not only to the state itself but to the nation as well. The cultural heritage, the national identity and national values were hence stressed as much as the sovereignty and integrity of the state. Notably, the emphasis put on the well-being and prosperity of the individual members of the nation was not limited to citizens of the state but included members living in diaspora outside the territory of the state. The second group, consisting of the three Baltic States of Estonia, Latvia and Lithuania, related survival purely to the state itself and hence foremost to territorial integrity. Notably, the fears related to the challenges of survival included the military threat not only from Russia but also from the Russian minorities living in each of the three countries.

Arguably, Czechia initially gave priority only to *influence*. However, following the Russian military aggression against Ukraine and the illegal annexation of Crimea, the Czech government has been giving as much priority to *survival*. While the economic dimension of security initially was the focus, the military dimension has received as much attention lately. Notably, Germany, that is the Republic's most important trade partner, is also the ally that the government is putting the greatest emphasis on in developing closer military cooperation within the NATO framework.

We find that three states, Hungary, Poland and Slovenia, have given as much priority to *survival* as to *status*. However, the focus of the three states differs. In the Hungarian case, the government gave attention not only to the state itself regarding survival but also to the Hungarian nation as well. Specific attention was hence given to the situation of the ethnic Hungarians living outside Hungary. Regarding status, the government put pride among the traditions of the Hungarian state and emphasised the need for gaining international recognition. The Polish government expressed similar considerations in this regard and perceived Poland as a prominent member of the Western community. However, contrary to

178 *Explaining the findings*

its Hungarian colleagues, the Polish government also perceived a clear and present danger from Russia; hence the need for focusing on survival as well. When it comes to Slovenia, we observe that the ends have been about the survival not only of the Slovenian state but of the nation as well. However, we also observe ends striving for international recognition.

Finally, regarding Romania, we observe that the government has expressed objectives indicating considerations on influence. Nevertheless, we conclude that *status* has been the prioritised end. Gaining recognition within NATO and the EU, especially from the key members of these organisations, being an important and credible member/partner has hence been the focus. Potentially, Romania may shift focus in order to increase its influence once the quest for status has been achieved. Notably, the Romanian government has also emphasised the sovereignty and integrity of the Romanian state, as well as the cultural and historical identity of the Romanian nation. Despite this observation, we argue that survival never has been the prioritised end.

15.3 The means element of the military strategy

When it comes to designing the means of the armed forces, we observe that one-half of the cases, that is Bulgaria, Croatia, Romania, Slovakia and Slovenia, gave priority to *expeditionary warfare*, while the remaining half, that is Czechia, Estonia, Hungary, Latvia, Lithuania and Poland, instead focused on the means for *national defence*.

Regarding the former category, we observe that only Romania really put deeds behind the words after articulating ambitions to transform the armed forces towards expeditionary capabilities. Notably, as in the Bulgarian case, the explicitly announced intentions regarding contributions to the international peace support operations decreased significantly once the memberships in NATO/EU were secured. In other cases such as Croatia and Slovenia, two of the services, that is the air force and the navy, have been rather unfit for contributing to expeditionary ambitions. Consequently, the army has had to bear the burden regarding participation in international military efforts. However, since the armies of these new allies often kept their old Russian-style equipment as surplus, the necessary funding for a more compelling transformation towards an expeditionary military has been delayed. The Slovakian government even admitted officially that the transformation of the Slovakian armed forces had been a failure in this regard. Although the Romanian armed forces also have been downsized, the capacity for national defence has remained at a high level. Notably, in the Romanian case, units from all three services have been deployed abroad including combat aircraft and frigates. At its peak, Romania had about 1,800 troops deployed to Afghanistan. Consequently, we argue that only Romania has been trustworthy when articulating expeditionary ambitions.

We find it reasonable to conclude that the Czech government has also put deeds behind the words regarding transforming the means towards a force capable conducting expeditionary warfare. However, since the war in Ukraine, we argue that the means for national defence have been prioritised. Moreover, the Czech ambitions to integrate with the German *Bundeswehr* regarding high-readiness and deployable forces for international missions has lately also included similar ambitions regarding more robust and heavy forces for collective defence operations. We conclude that this indicates increased attention on warfighting in the latter context. Contrary to its Czech neighbours, Poland has over time focused on national defence. Notably, this focus has not prevented Poland from providing impressive contributions to several and differently led international military operations. Arguably, the challenge for the government in each of the three Baltic States has been on balancing between the means essential for fighting the initial phase of a potential armed aggression solely by itself and the means necessary for receiving and hosting allied reinforcements. Regardless of the outcome of this balancing act, the focus has, at least in the Estonian and Latvian cases, always been on the national defence. Notably, all the Baltic States have expressed ambitions regarding expeditionary capacity. However, only Lithuania has contributed to international military operations with units from all three services. Following the 2014 Russian war against Ukraine, the Lithuanian focus has undoubtedly also been on national defence. Finally, regarding Hungary, we argue that the developments of the Hungarian armed forces have been less impressive. Three decades after the end of the Cold War, the means of the army still have a Russian profile. Moreover, the government tends to avoid referring to numbers when it comes to the means. Despite the explicitly declared pledge of sending 1,000 soldiers on international missions at any given time, the contributions to international military efforts have often been companies rather than battalions. Notably, in 2020 the Hungarian government admitted that no comprehensive, system-level development had taken place within the armed forces. Consequently, by neglecting to establish a trustworthy expeditionary capacity, we conclude that the focus of the armed forces, intentionally or not, has been on the national defence.

15.4 The ways element of the military strategy

When it comes to the final element of the military strategy, that is the ways, we separated the unilateral and multilateral approaches. Notably, this can have both a national and an international dimension. Regarding the later dimension, none of the 11 new European allies has articulated preferences for a unilateral approach. Consequently, when contributing to global peace and security, all new allies prefer taking part in multinational military formations conducting conflict prevention as well as other peace support operations in cooperation with other allies. The differences we have observed in this regard have been discussed when elaborating on the alignment strategy. Clearly, the new allies have slightly different preferences regarding who should lead these international operations. However, when it comes to the ways, the question is not related to whom but to whether or not to cooperate with others. The national dimension has two distinct outcomes. On the one hand, there is conducting

180 *Explaining the findings*

territorial defence operations in order to defend the home country against armed aggression. On the other is taking part in NATO collective defence operations outside the country's own territory. While external operations within the framework of collective defence by definition include a multilateral approach, the operations on domestic soil presents two options: either a unilateral or a multilateral approach. Regarding these options, some interesting differences have been observed. One such difference is whether the country has enough military resources to make a unilateral approach regarding defence operations on its own territory trustworthy or not. Arguably, the lack of means for projecting especially air and/or maritime power in the national context may make the multilateral approach a question not solely of preference but of necessity. Based on this elaboration, we find it reasonable to cluster the 11 cases in four categories.

The *first* category includes states that mainly focus on international peace support operations and seldom, if ever, elaborate on aspects of defending the home state and/or other allies. This category includes only Romania. At its peak, the country had about 1,800 troops deployed just to Afghanistan. Moreover, the armed forces have been deployed in different organisational contexts, hence putting deeds behind the multiple-courting alignment strategy. These deployments also make the multilateral approach trustworthy. The *second* category includes states that both focus on defending not only their own state but the territory of the allies as well and that, at the same time, are willing to contribute to international peace support operations. This category includes Czechia and Slovakia. Arguably, both these states have developed means making such an approach trustworthy. The *third* category includes states that lack the necessary resources to make a unilateral approach compelling in the national as well as in the international context. This category includes Bulgaria, Croatia, Estonia, Latvia, Lithuania and Slovenia. In all these cases, the lack of convincing national air power makes the multilateral approach a necessity rather than an option. In all but the Bulgarian cases, the lack of convincing resources also includes maritime means. Notably, despite the potential for contributing with ground forces to the defence of other allies' territory, Bulgaria has unilaterally limited the scope of such potential operations to counter- and anti-terrorism. With outdated means for the army and having to lease rather than procure combat aircraft for the air force, the capacity of the armed forces of Hungary indicate that this case also should be included in the third category. Although we observe tendencies towards developments intended to increase the capacity of the armed forces, we decide to do so. Since we also observe increased emphasis on political unilateralism, either of these two tendencies may very well lead to a shift in the Hungarian strategy toward a pure unilateral or a more balanced uni- and multilateral approach in the near future. The *fourth* category includes states with such a balanced approach. Currently, only Poland is included in this last category. By contributing with quite impressive resources to several different contexts regarding the use of force, as well as several bi- and multilateral arrangements regarding not only territorial defence but also the development of military power, the Polish government has indicated its preference for a multilateral approach. However, the government has also continuously expressed its preparedness for a unilateral approach if deemed necessary in the national context. We conclude that this is a central part of the Polish deterrence strategy and categorise Poland's design regarding the ways as a balance between the uni- and multilateral approaches.

16 Explaining the diversity of strategic responses

This chapter presents a cross-cases comparison focusing on how differences and similarities regarding our three intervening variables – that is (1) relative power and position in the international system, (2) national geographical characteristics and (3) historical experiences – covariate with the differences and similarities in the defence strategies. More specifically, we want to see if similarities related to a certain intervening variable among a subset of our cases produces similar strategic priorities regarding the choice of alignment strategy and/or prioritised military strategic ends, means and ways. Hence our ambition is to explain the diversity of strategy within the whole group of new allies. The influence of the intervening variables is analysed in three separate sections. These sections are followed by a fourth section in which the impact of the three intervening variables are summarised. In the fifth section, the focus returns to the dependent variable, that is the defence strategy. In this final section, we present our overarching conclusions regarding the influence of the intervening variables on the alignment strategy as well as on each of the elements of the military strategy.

16.1 Positional approach: differences in relative power

Our first intervening variable, relative power and position in the international system, relates to research on how power asymmetries between different categories of states forces comparably less resourceful states, such as middle powers and small states, to develop strategies and strategic priorities that are different from those pursued by great powers.[1] How and to what extent do differences in relative power covariate with differences in the defence strategies of the cases explored in this study?

For reasons elaborated on in Chapter 2, the 11 new allies are clustered into two categories, minor middle powers and small states. This categorisation is based on differences relating to access to latent economic, military and political power resources. To be included in the category of minor middle powers a state must meet the criteria of (1) being among the world's top 70 largest economies in terms of GDP and GDP per capita and (2) having among the top 70 largest accumulated military expenditures over the last 10 years. Since there is no 'minor middle power club' corresponding to the G20 for major middle powers, we have collected

DOI: 10.4324/9781003298052-19

Explaining the diversity of strategic responses 183

data on diplomatic representation measured as amount of foreign embassies in each state. To be recognised as a minor middle power, a state must (3) host a minimum 70 of foreign embassies. Together, these three indicators establish a 70–70–70 criterion. Four states, Czechia, Hungary, Poland and Romania reached this triple criterion and are hence considered to be minor middle powers. The remaining new allies – Bulgaria, Croatia, Estonia, Latvia, Lithuania, Slovakia and Slovenia – are categorised as small states. However, as mentioned in Chapter 2, two of our cases merit special concern. Poland has a great advantage to the other states regarding both economic and military capabilities. Arguably, this may motivate that Poland should be classified as a 'middle power' rather than as a 'minor middle power'. Slovakia, although narrowly, reached our criteria for minor middle powers in terms of economic and political power resources but did not reach the criteria for diplomatic representation. In the case of Poland, we will evaluate to which extent it pursues more ambitious strategies than other minor middle powers. If so, it may be reasonable to argue that Poland should be classified as a middle power. This subcategory of states is considered to be less able and resourceful than major middle powers but still capable of pursuing more ambitious strategies than minor middle powers. Regarding the final classification of Slovakia, we are interested to see whether its strategic priorities mostly correspond to those of small states of minor middle powers.

If differences in relative power among the 11 new allies covariate with differences in their alignment strategies and/or priorities regarding military strategic ends, means and ways, we have established a covariance that indicates the importance of considering differences in relative power in analyses of states' strategic choices. However, to strengthen our case for a causal link between the intervening variable and the dependent variable, we will also present an argument on how differences in relative power are likely to affect states' strategic priorities. Additionally, the different intervening variables may also interact and demand a simultaneous presence of two or more variables to produce a specific strategic response.

16.1.1 Alignment strategies

Since nine of our cases, including six small states and three minor middle powers, pursued some kind of *multiple-courting* strategy, the distinction between minor middle powers and small states does not produce a covariation between differences in relative power and the choice of alignment strategy. Moreover, both the military-oriented and the broader security-oriented approach to multiple-courting include both small states and minor middle powers with Bulgaria, Estonia, Latvia, Lithuania and Romania favouring the former and Croatia, Czechia, Hungary and Slovakia preferring the latter. Notably, Poland, the most resourceful state, follows its own path by pursuing an *offensive bandwagoning* strategy towards the US including its substantial contribution to the invasion of Iraq 2003. Arguably, this indicates that more resourceful categories of states may have alignment options open to them that less resourceful states have not. Slovenia's modest contributions

184 *Explaining the findings*

to allied collective efforts and its dependency on NATO funding for developing military infrastructure may be related to a lack of military capacities. However, since the other small states did not pursue *passive bandwagoning*, the Slovenian case does not help us in establishing a covariance between relative power and the choice of alignment strategy.

16.1.2 Strategic ends: survival, influence and/or status?

Ten of the new allies gave, in one way or another, priority to *survival*. The only exception was Romania that prioritised *status*. Six of the cases, Bulgaria, Croatia, Estonia, Latvia, Lithuania and Slovakia, focused solely on survival, while Czechia, Hungary, Poland and Slovenia put the emphasis on survival in addition to one of the other two main ends. The six states focusing solely on survival are all categorised as small states. In the second group, all states but Slovenia are categorised as minor middle powers. Czechia initially gave priority solely to *influence*. However, following the Russian military aggression against Ukraine and the illegal annexation of Crimea, the Czech government has put as much emphasis on survival. We find that three states, Hungary, Poland and Slovenia, have given as much priority to survival as to *status*. This indicates a general tendency that increased size correlates with increased ambitions regarding ends related to influence and status among minor middle powers and small states.

The six small states that focused solely on survival explicitly emphasised their priority by presenting its *territorial integrity* as the 'primary goal' (Bulgaria) or a 'vital interest' (Bulgaria, Croatia, Lithuania and Slovakia) in contrast to other 'important' interests. Others within this group formulated this priority as the 'main task' of the armed forces (Estonia) or an 'overarching aim of the defence policy' (Latvia). Within this group, the emphasis on survival of Bulgaria, Croatia and Slovakia, concerned both the state itself and the nation's cultural heritage, identity and values.

The four states that prioritised survival in addition to influence or status presented the defence of security as one of several core objectives. Examples of complementary objectives within this second group were establishing a 'reputation as a trustworthy and reliable ally' (Czechia), gaining international prestige and influence (Hungary and Poland) and 'a strong international position' (Poland) or reputation and national identity (Slovenia). For Romania, its interest in promoting status concerned its position as a NATO as well as a EU member and finding support for Romania's security, economic and political development. Similar to some of the states in the first group, the governments of Hungary, Poland and Slovenia emphasised the need for protecting national identity, values and, occasionally, a diaspora living in neighbouring states.

16.1.3 Military means: national defence and/or expeditionary warfare?

When it comes to designing the means of the armed forces, we observe that one-half of the cases, that is Bulgaria, Croatia, Romania, Slovakia and Slovenia, gave priority to *expeditionary warfare*, while the remaining half, that is Czechia, Estonia, Hungary, Latvia, Lithuania and Poland, focused on the means for *national defence*. Again, we find that minor middle powers and small states appear in both subcategories of our operationalisation of the dependent variable. However, a closer look at the extent to which these states have actually transformed their defences, in order to be able to contribute to common efforts related to national defence and expeditionary warfare, provides a more nuanced picture.

Regarding expeditionary warfare, Romania – the only minor middle power among the group of states prioritising expeditionary warfare – is the only state that really put deeds behind the words in transforming the armed forces towards expeditionary capabilities. Romania has deployed units from all three services abroad, including combat aircraft and frigates, and had at its peak about 1,800 troops deployed to Afghanistan. In the Bulgarian case, its announced intentions regarding contributions to the international PSOs decreased after its memberships in NATO/EU were secured. Croatia and Slovenia, two of the small states prioritising expeditionary warfare, let their army bear the burden of providing contributions to international PSOs. Additionally, the preservation of old USSR-era equipment as surplus among these states further delayed the transformation towards an expeditionary warfare-oriented military. The Slovakian government officially admitted that the transformation of the Slovakian armed forces towards expeditionary warfare had been a failure. Hence our results regarding this group of states indicates that minor middle powers seem to be more able to put actions behind their words when developing military capacities related to expeditionary warfare.

The claim that minor middle powers seem to be relatively more capable of developing capacities for expeditionary warfare finds further support in the fact that the Czech government, motivated by the original strategic aim of increasing its influence, initially transformed its military towards a force capable conducting expeditionary warfare. However, Russia's illegal war against Ukraine in 2014 changed the Czech government's priorities, now focusing on means related to national defence. The government's new priorities is further reflected in its changing ambitions regarding the cooperation with Germany and the German *Bundeswehr*. Initially, this cooperation concerned the creation of high-readiness and deployable forces for international missions. After 2014, this cooperation was refocused on creating more robust and heavy forces for collective defence operations. Contrary to Czechia, Poland has consistently focused on national defence. However, Poland has simultaneously been able to provide impressive contributions to several international military operations. The three Baltic States have faced the common challenge of balancing between the means essential for fighting alone during the initial phase of an armed conflict and developing capacities necessary for receiving and hosting allied military support. In the Estonian and Latvian cases, the general orientation of the means have consistently been on national defence. Lithuania's comparably ambitious attempt to contribute to multilateral PSOs with units from all three services before 2014 indicate that even small states may have greater ambitions in regard to expeditionary warfare.

186 *Explaining the findings*

However, following Russia's 2014 war against Ukraine, Lithuania too refocused on national defence.

Compared to the other three minor middle powers, Hungary has been less inclined to transform its armed forces in any direction. It has kept the old USSR profile of its armed forces and officially admitted that no comprehensive system-level development has taken place. Therefore, we conclude that its focus remains on national defence. The Czech and the Polish governments' more ambitious responses to systematic pressures indicate that minor middle powers have options regarding *both* national defence and expeditionary warfare that in most cases is not open to less resourceful small states. However, the unwillingness of the Hungarian government to transform its armed forces indicates that it is necessary to consider the potential influence of other intervening variables as well.

16.1.4 Ways: unilateral and/or multilateral approach?

Our analysis of ways departs from a distinction between unilateral and multilateral approaches. Both approaches may have national as well as international dimensions. None of the new European allies has articulated preferences for a unilateral approach regarding the international dimension. This is not surprising considering that our selection of cases is limited to minor middle powers and small states. The national dimension have two distinct outcomes: conducting territorial defence operations in order to defend the own country against armed aggression and contributions to NATO collective defence operations outside the country's own territory. While collective defence by definition includes a multilateral approach, the operations on domestic soil can be pursued with both unilateral and multilateral approaches.

Based on these distinctions, the strategic priorities of the 11 new allies are divided into four categories. The first category includes states that *mainly focus on international PSOs* and seldom elaborate on aspects of defending their own states and/or other allies. This category includes only Romania, which made substantial military contributions to different organisational contexts, hence making its multilateral approach trustworthy. As further discussed in the next section, Romania has also expressed ambitions developing unilateral capacities to deter and counter aggression against its territory. The second category includes states that both focus on *defending the own state* and *the territory of the allies* as well and at the same time are willing to contribute to international PSOs. This category includes the minor middle powers of Czechia and Slovakia, the latter of which has been preliminarily classified as a small state. When it comes to the ways, Slovakia signals ambitions of aiming to punch above its small state weight. However, huge challenges in Slovakia's defence transformation process, related to understaffed professional forces, failed armaments projects and a low level of interoperability, do not strengthen its case for being considered as a minor middle power. The third category includes states with a *balanced approach* using both unilateral and multilateral approaches. Currently, only Poland is included in this category. By contributing with quite impressive resources to several different contexts regarding both the use of military force and the development of military power, the Polish government has indicated its preference for the multilateral approach. However, the government has also continuously expressed its preparedness for a unilateral approach if deemed necessary in the national context. Arguably, this approach would not be an alternative if Poland did not have power resources do make such a strategy credible. Poland's greater ambitions when it comes to both means and ways support the argument that Poland should be considered a 'middle power' in a class of its own among the new allies.

The fourth category includes states that *lack the necessary resources to make a unilateral approach compelling* in the national as well as in the international context. All states categorised as small states, except Slovakia, belong to this category, indicating a strong correlation between relative power and ability to develop credible unilateral capacities. In all these cases, the lack of convincing national air power makes the multilateral approach a necessity rather than an option. In all but the Bulgarian cases, the lack of convincing resources also includes maritime means. Hungary is difficult to squeeze into any of the four categories. With outdated means for the army and having to lease rather than procure combat aircraft for the air force, the capacity of the armed forces of Hungary indicates that this case also should be included in the fourth category. However, we have recently also noted tendencies towards developments intended to increase the capacity of the armed forces and increased emphasis on political unilateralism. It remains to be seen how this development is affected by Russia's invasion of Ukraine in 2022.

In aggregating how differences in relative power covariate with priorities regarding strategic ways, small states' lack of resources seems to covariate with the lack of convincing ambitions with regard to unilateral approaches. Slovakia provides an exception to this pattern. However, its limited military capabilities and capacities do not strengthen its case for being included in the category of minor middle powers. When it comes to this category, we see a less coherent picture regarding prioritised ways. This indicates that relative power and position in the international system is not enough to affect the strategic priorities regarding ways, at least not without additional support from other intervening variables.

16.1.5 Conclusions regarding the positional approach

In evaluating the positional approach, we find that differences in power resources between minor middle powers and small states did not covariate with strategic priorities regarding the external efforts and *alignment strategies*. However, regarding the military strategy and the internal efforts, the cross-case comparison revealed a pattern regarding strategic priorities indicating similarities within the two categories and differences between them.

Regarding *ends*, the small states included in this study all gave priority to objectives related to survival. With the exception of Romania, the minor middle powers all prioritised survival in addition to either influence or status. Arguably, our findings indicate a general tendency that increased power correlates with increased ambitions.

188 *Explaining the findings*

Regarding the *means* and the *ways*, all minor middle powers, except Hungary, proved to be willing and able to put deeds behind their words when it came to transforming their armed forces. They all, except Hungary, developed the capacities needed for both national defence and expeditionary warfare and also provided substantial contributions to international PSOs. None of the states classified as small states were able to present convincing ambitions in this regard. We find indications of a recent increase in the Hungarian government's ambitions when it comes to modernising the armed forces. However, the difference between Czechia's and Hungary's defence strategies during the main part of the first two decades of the twenty-first century seems difficult to explain with reference to position in the international system.

16.2 The geographical approach

Our second intervening variable, national geographical characteristics, is related to research on geopolitics and military strategy. The Swedish political scientist Rudolf Kjellén, who coined the word 'geopolitics', used it to describe the sum of a particular state's geographical characteristics, its natural endowment and resources. During the twentieth century, the concept became an integrated part of the strategies of great powers. Examples of this are Nazi-Germany's eastward expansion, US containment policy against world communism and the USSR and the Brezhnev doctrine. These competing Cold War visions of the world order was replaced in the early 1990s by George Bush's conceptualisations of a new US-led liberal world order and a 'Europe whole and free' (Bush 1991; see also Brezhnev 2006; Truman 2006; Tuathail 2006).

The 11 states included in this study were mostly on the receiving end of these competing geopolitical conceptions and the different great power conflicts associated with them. The Munich Agreement of 1938, the Molotov-Ribbentrop Pact of 1939 and the Yalta Agreement of 1945 constitute, according to Ainius Lašas, a 'black trinity' that affected all 11 new European allies (Lašas 2010). Russia's invasion of Ukraine in February 2022, as well as renewed Russian demands for a sphere of influence encompassing former USSR republics and former members of the WP, can be interpreted as efforts to restore the twentieth century's geopolitical divisions of Europe and a return to the old European security dynamics discussed in Chapter 2.

Both classical realist scholars, structural realists and researchers within the field of Strategic Studies have acknowledged the importance of geography in explanations of both outcomes of conflicts and priorities regarding defence planning (Gray 2006; Morgenthau 2006; Layne 2012; Mearsheimer 2018). According to Colin Gray, geography 'explains more about a polity's national security issues than does any other factor' (Gray 2015:84). Due to the great number of cases, we have focused on one central geographical aspect, that is a shared land border with Russia. If this intervening variable should prove to be of importance, we expect that the four states with a land border to Russia – Estonia, Latvia, Lithuania and Poland – will give greater priority to (1) cooperative alignment strategies,

(2) military means related to national defence and (3) the multilateral approach regarding the ways, compared to the states with a less geographically exposed position. In addition, we expect these four states to respond more firmly to the deteriorating regional security after 2008 and especially after 2014.

16.2.1 Alignment strategies

Regarding alignment strategy, the three Baltic States all prioritise a military-oriented approach to *multiple-courting*, complementing existing defence and security cooperation within the EU and NATO with bilateral arrangements with the US and other states such as Poland. The Baltic States also initiated a trilateral cooperation among themselves. The alignment strategies of these states are examples of external efforts to coordinate the use of military force with likeminded states in order to compensate for a lack of national military resources. Poland also prioritises its strategic partnership with the US, including active participation in the US ballistic missile defence program. Similar to the Baltic States, Poland has received allied troops on its soil. However, Poland's access to greater military resources has also made it possible for the Polish government to provide greater contributions to US-led coalitions, hence pursuing an *offensive bandwagoning* strategy.

Slovenia and the four states practicing the broader security-oriented approach all enjoy less geographically exposed positions. Consequently, geographical position seems to provide a strong covariance regarding different priorities in multiple-courting strategies. However, two additional states prioritise a military approach to multiple-courting: Bulgaria and Romania. The alignment strategies of these two states may indicate that the geographical aspect of strategic exposure should include additional, potentially troublesome aspects of a particular state's geographical position. However, Romania's priorities regarding ends and ways do not indicate that its choice of alignment strategy has been primarily motivated by perceived threats against its territory.

16.2.2 Strategic ends: survival, influence and/or status?

Regarding ends, the three Baltic States constitute a separate group among the new allies that relate *survival* primarily to the territorial integrity of the state and concerns related to Russian minorities. Poland, being a more resourceful state, has higher ambitions prioritising both survival and *status*.

Less geographically exposed small states such as Bulgaria, Croatia and Slovakia, related *survival* to both the state itself and the nation. Like Poland, Hungary and Slovenia practice a balanced approach prioritising both survival and *status*. Romania is the only state that does not prioritise survival; its priority is instead status and improvement of living conditions within Romania, indicating less concern with geographical exposure. Czechia's increased priority of survival following the Russian military aggression against Ukraine 2014, contradicts the expectation that states sharing land borders with Russia should respond more firmly to a more assertive and threatening Russia.

16.2.3 Military means: national defence and/or expeditionary warfare?

When it comes to the military means, all four states with land borders to Russia have been reluctant to dismantle capacities related to *national defence*. Moreover, they all responded to Russia's war against Ukraine in 2014 by increasing their efforts to develop military capabilities needed to defend themselves against a qualified opponent. However, there are significant differences in the strategic priorities related to national defence among the three small states as well as between them and Poland.

Among the three small states, Estonia has been most consistent and ambitious in prioritising the development of capacities related for national and collective defence. Its military expenditures as a percentage of GDP was in 2020 the highest of all 11 new allies, at 1.92 percent (see Chapter 2). In 2009, well before Russia's war against Ukraine, it prepared itself both to respond to a sudden attack with national means and to receive allied forces by air, land and sea by investing in air defence, mine clearance and preparing the defence of strategically important areas. The main priority of the land forces was initially to create one high-readiness brigade consisting of a professional battalion supported by conscript units. In 2013, the Estonian MoD declared its ambitions to have one additional infantry brigade operational by 2022. Four years later, the MoD declared its ambitions to enlarge Estonia's wartime rapid response structure from 21,000 troops to 25,000 and to increase the number of conscripts. Regarding combat and fighting vehicles, Estonia has since 2000 increased its park of ACVs, APCs and IFVs from 32 to 180. When it comes to the navy, the Estonian government has prioritised high-speed patrol vessels and mine hunting vessels. Estonia lacks, on the one hand, combat aircraft but has, on the other hand, invested heavily in developing military cyber defence capabilities.

Similar to Estonia, Latvia lacks combat aircrafts, and its navy is focused on tasks related to mine clearance, surveillance, as well as search and rescue operations. In addition, both countries prioritises air defence and anti-tank capabilities. However, regarding land forces, the two countries have slightly different ambitions and priorities. In 2004, Latvia declared its ambition to create a fully professionalised force and put greater emphasis on contributing to PSOs, even though the size of their contributions were rather modest. Moreover, the Latvian government declared that its priority was the quality of its forces, not their numbers. In 2020, Latvia's defence expenditures were 1.30 percent of GDP despite previous ambitions to reach 2 percent by this year (see Chapter 2). In 2016, the two National Guard regions were referred to as brigades, but according to IISS's statistics, Latvia had only one brigade operational in 2020. Moreover, Latvia's park of military combat and fighting vehicles dropped from 13 in 2000 to zero in 2020. Altogether, this suggests that while both Estonia and Latvia prioritise national defence against a qualified adversary, Latvia's efforts in this regard are less consistent and ambitious.

The armed forces of Lithuania have a force posture similar to the other two Baltic States, focusing on military capabilities for national and collective defence.

However, prior to Russia's attack on Ukraine in 2014, Lithuania had higher ambitions regarding contributions to international PSOs compared to the other two. In 2012, the government announced its ambition, having 50 percent of the land forces prepared to be deployed outside its territory. The ambitions included contributing to NATO-led operations with up to one infantry battalion battle group, one special operations forces squadron, one MCMV ship and one light transport aircraft. Additionally, the government declared its willingness to provide minor contributions to EU- and UN-led missions. However, following Russia's war on Ukraine in 2014, Lithuania refocused its efforts towards national and collective defence. Regarding external efforts, the Lithuanian government strived for enhancing US military presence in the Baltic Sea region, and it cultivated various forms of military cooperation with neighbouring states. In 2015, the government announced its ambition reorganising the land forces, hence having one mechanised brigade and one motorised brigade operational. Two years later, in order to increase the size of the armed forces, the Lithuanian Parliament presented a mixed model consisting of professional military soldiers, conscripts and national defence volunteers. Moreover, the government organised a national rapid response force including new infantry and artillery battalions within the second brigade. Additionally, Lithuania has increased its number of combat and fighting vehicles from 14 in 2000 to 230 in 2020.

Prior to the Russian–Georgian War in 2008, the Polish government announced its ambitions to develop capacities for both *national defence* and *expeditionary warfare*. Regarding national defence, Poland's armed forces were to be capable of defending Polish territory both independently and as a part of NATO's collective defence. In 2009, the government stressed that the aim of the ongoing defence transformation was to improve the armed forces' ability to deter aggression and defend Polish territory. Similar to the other states with land borders to Russia, Poland has focused, from the Russian assault on Georgia and onwards, on developing military capacities related to defence against a qualified opponent. However, being a more resourceful middle power, Poland has been able to do this with higher ambitions. The core of its land forces is composed of four armoured and mechanized divisions. In addition, Poland has invested in air transport capacities to facilitate rapid deployment across its territory. Moreover, the Polish government has prioritised the modernisation of the air defence system, including missile defence. In 2018, Poland signed an agreement worth 4.75 billon USD for the Patriot missile defence system. An additional crucial element of the Polish deterrence strategy is long-range precision weapons, and in 2014, Poland was the first NATO country allowed to purchase the US Joint Air-to-Surface Standoff Missile system. Moreover, in 2019, the US approved the sale of 32 F-35 combat aircraft. Regarding the balance between the different services of the armed forces, Poland clearly prioritises land forces. However, in addition to the investment in new combat aircraft, Poland has also invested in its naval forces. Poland has also allocated comparably large economic resources to the armed forces. According to SIPRI, Poland spends 1.90 percent of GDP on military expenditures in 2021, resulting in an annual military expenditure of 10 billion USD. This more than equals the collective military expenditures of Bulgaria, Croatia, Czechia, Hungary, Romania and Slovakia (see Chapter 2).

Five of the seven new allies without land border to Russia, that is Bulgaria, Croatia, Romania, Slovakia and Slovenia, gave priority to *expeditionary warfare*, and one additional state, Czechia, initially made similar priorities regarding the primary design of its military means. Accordingly, during the early part of the twenty-first century, Hungary was the only exception to this pattern of covariance between having a land border to Russia and general priorities regarding means. Indicators of this are the reductions among these states regarding land forces, MBTs, combat aircraft and other advanced military systems needed in an armed conflict with a qualified state adversary. Another general tendency within this group is the ambition to transform their armed forces into professional volunteer forces able to contribute to international PSOs.

Two of the states in this group, Czechia and Romania, have higher ambitions when it comes to national defence as well. In 2017, when responding to a more assertive Russia and its armed aggression against Ukraine, the Czech government announced its intentions assigning a brigade to a German army division earmarked as follow-on forces for collective defence operations. The government also presented its ambitions of increasing its armed forces with an additional 5,000 military professionals. In 2019, the government prioritised the build-up of a heavy brigade and announced its decision to either extend the lease or acquire new combat aircraft no later than 2025. In parallel to its substantial contributions to various international operations, the Romanian government in 2015 prioritised the development of unilateral military capabilities to deter and counter a possible aggression against Romania until receiving assistance from allied forces. To achieve this end, Romania has two army division headquarters, 10 independent army brigades, 400 MBTs and substantial marine forces including three destroyers, four corvettes and 14 offshore missile and patrol craft. In 2020, the government announced its ambitions of establishing several installations on Romanian soil in order to strengthen NATO's enhanced forward presence.

16.2.4 Ways: unilateral and/or multilateral approach?

Regarding ways, the three Baltic States have little choice but to rely on *multilateral* strategies for both national defence and contributions to international PSOs. Poland, on the other hand, has greater *unilateral* ambitions when it comes to deterring Russia with its own military resources. Arguably, the Baltic States' geographically exposed position still necessitates the allocation of nationally controlled military resources, such as minesweepers and base defence units in order to facilitate allied support.

However, as mentioned in the previous section, preferences for multilateral strategies in relation to both national defence and expeditionary warfare are not unique for states with borders to Russia. Consequently, geographical position does not appear to covariate with priorities regarding unilateral and multilateral strategies. Some states with a less geographically exposed position tend to be more willing to contribute to collective defence outside their own territories, and some of them have greater ambitions regarding expeditionary warfare. Moreover, Poland has made substantial contributions to allied efforts outside Europe in spite of its geographically exposed position. The ability and willingness to develop relevant military capacities therefore seem to be the most important factor in explaining priorities regarding strategic ways.

16.2.5 Conclusions regarding the geographical approach

Regarding *alignment strategies*, differences related to geographical position covariates with different priorities regarding military-oriented multiple-courting strategies and broader security-oriented approaches. However, the alignment strategies of Bulgaria and Romania indicate that the geographical aspect of strategic exposure may benefit from including additional, potentially troublesome aspects of individual states' geographical positions. Moreover, Russia's invasion of Ukraine in 2022 is likely to increase the number of states that consider themselves strategically exposed due to their geographical position. For the four states with borders to Ukraine – Hungary, Poland, Romania and Slovakia – war has literally come closer to their borders, a sudden and probably unexpected change in their external security environment that especially Hungary is ill prepared for when it comes to military means.

When it comes to the *ends*, the cross-cases comparison revealed similarities between the three Baltic States' priority of survival and territorial integrity. Poland, being a more resourceful state, prioritised both survival and status. However, since survival was a prioritised end for most of the new allies, this is not a unique priority for states with a land border to Russia.

The analysis of priorities related to military *means* revealed that states with a land border to Russia have prioritised military capabilities related to defence against a qualified state actor. Initially, Lithuania had a comparably balanced approach in prioritising between means for national defence and expeditionary warfare. However, following Russia's attack against Ukraine in 2014, Lithuania refocused its priorities to national and collective defence. Before this attack, all states without land borders to Russia, except Hungary, prioritised military means and capabilities related to expeditionary warfare. As a part of their defence transformation processes, these less geographically exposed states dramatically reduced the size of their armed forces and weapon systems related to national defence against a qualified adversary. However, following the Russian aggression against Ukraine in 2014, Czechia refocused on national defence and deepened defence cooperation with allied states.

Our analysis of *ways* indicates that national differences in access to relevant military means is the most important national characteristic when explaining the choices between unilateral and multilateral approaches related to both national defence and expeditionary warfare.

194 *Explaining the findings*

16.3 The historical approach

The influence of previous experiences of armed conflicts on states' strategic priorities is well documented (Snyder 1991; Levy 1994; Kier 1997). Experiences received during 'systematic' great power wars have, according to Dan Reiter, the greatest impact on future choices between alternative alignment strategies (Reiter 1994). For the reasons presented in Chapter 2, the covariation between experiences of armed conflicts and strategic priorities regarding defence strategies are analysed along three dimensions and focus on experiences received during the last great power war, that is WWII. Arguably, it is reasonable to assume that experiences from the latest systematic war have the greatest impact on post-war strategic priorities. The first dimension concerns experiences of national defence against aggression and hence especially from great powers, the second dimension experiences of offensive warfare in alliance with Nazi-Germany and/or the USSR, and the third experiences of military assistance from and/or cooperation with one or several Western great powers.

Based on the 11 new allies' different experiences of national defence during WWII, we have divided them into two groups. The first group consists of the states that during the war ceased to exist as political units and whose territory was completely annexed and/or controlled by one or several foreign powers. The members of this category are Czechia, Estonia, Latvia, Lithuania, Poland and Slovenia. The second category consists of the states that were able to keep at least some form of independence during WWII. While Croatia and Slovakia were established as German puppet states, Bulgaria, Hungary and Romania kept their independence. Notably, all five states in this category fought with Nazi-Germany in the initial phases of WWII and were occupied by the USSR during the final phase.

Regarding the experiences of national defence, we expect that states that ceased to exist will pursue hedging strategies along the lines of the military-oriented approach to multiple-courting. We also expect them to give higher priority to ends related to survival and territorial integrity. Moreover, we expect that they will be more reluctant to dismantle their capacities for national defence and more unwilling to transform their armed forces towards expeditionary warfare. When considered possible, these states are expected to give higher priority to unilateral strategies to national defence. For the second group, our expectations are the opposite: hedging strategies focusing on broader security-oriented approaches, lower priority given to ends related to territorial security, a greater preparedness to transform their armed forces towards expeditionary warfare and the pursuit of multilateral ways to share and integrate their military resources with other states.

Wartime experiences related to the three dimensions also concern good and/or bad experiences related to specific great powers. Are states that were invaded and occupied by the USSR responding more harshly to a renewed potential Russian military threat compared to states that were not fighting the USSR? Do states that suffered attacks from Nazi-Germany find it more difficult to cooperate with the EU and/or NATO in matters related to collective defence? Are states with positive

Explaining the diversity of strategic responses 195

or negative experiences from joining forces with Western great powers more or less inclined to commit themselves to allied war efforts?

In analysing differences related to historical experiences, we also, as mentioned in Chapter 2, consider the time of lasting independence prior to WWII and during the Cold War as a main dividing line regarding historical experience related to state-building. Consequently, the four states that gained independence prior to WWII and continued to exist as states during the Cold War era – Bulgaria, Hungary, Poland and Romania – are regarded as a group of states that, due to their longer history, are less likely to perceive threats to their future existence and independence. Arguably, Czechia can be included as an additional member of this group, depending on whether it is interpreted as a continuation of the former Czechoslovakia or not. Regarding the six other new allies, the three Baltic States lost their independence during WWII and experienced continued annexation by the USSR during the Cold War era. Croatia and Slovenia gained their independence only after the end of the Cold War following the violent dissolution of Yugoslavia, whereas Slovakia gained independence following the peaceful dissolution of Czechoslovakia. Presumably, the latter group will give greater priority to ends related to protecting their newly gained independence, while the former group may also prioritise increasing their influence or status. Differences related to experiences of national independence may also affect the views on the means, with the latter category focusing on national defence and the former on using their military expeditionary in order to gain status and/or influence.

16.3.1 Alignment strategies

Regarding alignment strategies, our cross-cases comparison focuses on the 11 new allies' experiences of national defence, offensive warfare together with great powers, and experiences of cooperating with and/or receiving military assistance from one or several Western great powers during WWII. In Table 16.1, the experiences of the new allies are summarised.

Regarding the group of six states whose territories were completely annexed/controlled by other states, the three of Baltic States all prefer the *military-oriented approach* to *multiple-courting*. A fourth state, Czechia, initially prioritised a broader security approached. Arguably, this can be seen as a gradual move towards a more military-focused approach as it from 2014 and onwards intensified its defence cooperation with neighbouring states such as Germany. However, since the Czech parts of Czechoslovakia were annexed by Germany, Czechia's choice of partner does not correspond well with its WWII historical experiences. Poland, a fifth state in this group, is considered as pursuing an *offensive bandwagoning strategy* towards the US. This strategy has increasingly focused on military cooperation and can therefore be seen as consistent with a perceived need to prioritise national and collective defence. The last member of this group, Slovenia, deviates from our expectations regarding annexed states, practising a *passive bandwagoning strategy*.

Table 16.1 WWII experiences of armed aggression.

	National defence against aggression, especially from great powers	Offensive warfare (OW) with Germany and/or the USSR	Military assistance (MA) and/or military cooperation (MC) with Western great powers
Bulgaria	Attacked and occupied by the USSR in 1944.	Initially OW with GERMANY against Greece and Yugoslavia.	
Croatia	Invaded by GERMANY et al. in 1941. Parts of territory annexed by Italy and Hungary. Remaining parts a German puppet state.		Support by the UK to Yugoslavian resistance.
Czechia	Annexation of Sudetenland in 1938. Invaded by and loss of national independence to GERMANY in 1939; continued resistance.		MC with the UK et al.
Estonia	Invaded and annexed by the USSR in 1940, thereafter occupied by GERMANY 1941–1944. Renewed annexation by the USSR from 1944.		
Hungary	Invaded and occupied by GERMANY in 1944 and by the USSR 1945.	Initially OW with GERMANY against several neighbouring countries, including the USSR.	
Latvia	Invaded and annexed by the USSR in 1940; thereafter occupied by GERMANY 1941–1944. Renewed annexation by the USSR from 1944.		
Lithuania	Invaded and annexed by the USSR in 1940; thereafter occupied by GERMANY 1941–1944. Renewed annexation by the USSR from 1944.		
Poland	Invaded, partly annexed and occupied by GERMANY and the USSR 1939–1941. From June 1941 only by GERMANY; continued resistance.		MC with the UK et al.

	National defence against aggression, especially from great powers	Offensive warfare (OW) with Germany and/or the USSR	Military assistance (MA) and/or military cooperation (MC) with Western great powers
Romania	Part of territory annexed by Hungary in 1938 and by the USSR in 1940.	OW with GERMANY 1941–1944. OW with the USSR 1944–1945.	
Slovakia	Separated from Czechia by GERMANY in 1939 as a German puppet state. Part of territory annexed by Poland and Hungary before the WWII.	OW with GERMANY.	
Slovenia	Invaded and annexed by GERMANY et al. in 1941.		

However, the group of states focusing on military approaches to multiple-courting includes two states that were not annexed during WWII, that is Bulgaria and Rumania. Arguably, this indicates that this approach is not unique to states that suffered annexation.

Regarding relationships with specific great powers, Poland shares Czechia's experience being invaded and annexed by Germany. Moreover, similar to Czechia, Poland has deepened its defence cooperation with Germany after joining NATO. It appears that Germany, in contrast to the USSR/Russia, has managed to break with the past and convince the new allies that it is an asset rather than a threat to their security. Naturally, Germany's membership in both NATO and the EU has contributed to the makeover of the traditional threat perceptions regarding Germany. Obviously, it is difficult to find any clear pattern of covariance regarding alignment strategies among the states that, on the one hand, pursued offensive warfare with Germany and/or the USSR and that, on the other hand, cooperated and/or received assistance from Western powers during WWII. Our results indicate that the WWII experiences are not as important to the strategic priorities regarding alignment strategy as expected in previous research on the lasting impact of systematic wars (see, for example, Reiter 1994; Edström *et al.* 2019). Potentially, the legacy of WWII has been overshadowed by the 11 new allies' experiences during the Cold War era. Arguably, Czechia's increased priority of collective defence with NATO partners after 2014 may be motivated, at least partly, by the fact that it suffered an armed intervention from the USSR and other WP countries in 1968. Additionally, common membership in the EU and NATO seems to have facilitated cooperation between states involved in existential struggles during WWII.

16.3.2 Strategic ends: survival, influence and/or status?

When it comes to strategic ends, the three Baltic States all prioritise *survival* related to territorial security, as expected by states suffering annexation. Following

Russia's military aggression against Ukraine in 2014, Czechia seem to have refocused on survival instead of influence. Poland, being the most resourceful state among the new allies, prioritises both survival and status. Slovenia, possibly due to its less geographically exposed position, again deviates from our expectations regarding annex states by prioritising both survival and status.

Regarding the partly overlapping approach to historical experiences that focus on lasting independence before WWII and during the Cold War era, the 11 new allies are divided into two groups. The first group consists of the four states that gained their independence before WWII and continued to exist as states during the Cold War: Bulgaria, Hungary, Poland and Romania. Czechia is considered as an additional fifth member of this group since it can be viewed as the heir of Czechoslovakia. The second group includes the three Baltic States that were annexed by the USSR during WWII and the three new allies that gained their independence only after the end of the Cold War, that is Croatia, Slovakia and Slovenia. Five of the six states in the second group focused on survival. Slovenia again breaks the pattern by prioritising both survival and status. Three of them, the Baltic States, focused specifically on territorial integrity. All six states had parts of their territory annexed during WWII. However, in the case of Croatia, Slovakia and Slovenia, it was Nazi-Germany and its allies that were responsible for these aggressions. For the Baltic States, the USSR was the main aggressor, and Russia – in contrast to Germany – continues to challenge their independence.

The group of states with continued independence during the Cold War were more inclined to prioritise additional strategic ends such as *status* and *influence*. However, this group includes all the comparably resourceful states classified as minor middle powers and Poland.

16.3.3 Military means: national defence and/or expeditionary warfare?

Historical experiences of lasting independence does not covariate with present priorities regarding national defence and expeditionary warfare. Two of five states belonging to this group, Poland and Hungary, prioritised national defence. Bulgaria and Romania prioritised expeditionary warfare, and the fifth state, Czechia, prioritised expeditionary warfare until 2014 when the Czech government changed preferences.

However, there is a relatively strong covariance between, on the one hand, experiences of complete annexation and/or loss of independence during WWII and, on the other hand, prioritisation of developing and preserving military capacities needed to fight a qualified state aggressor. Among the six states belonging to this group, the three Baltic States, Poland and Czechia clearly prioritised military means related to national defence after 2014. The only state that did not follow this pattern was Slovenia. Moreover, among the states that did not suffer from total annexation and/or loss of independence, only Hungary has continued to prioritise national defence when it comes to military means. Additionally, as previously mentioned, our reasons for classifying Hungary as a state that give priority to national defence are primarily related to the country's lack of results

Explaining the diversity of strategic responses 199

in transforming its armed forces in any direction. Contrary to Hungary, all other states that escaped total annexation and/or losing their independence have been more prepared to develop military capabilities and capacities needed for expeditionary warfare.

16.3.4 Ways: unilateral and/or multilateral approach?

Regarding the priorities on multilateral and unilateral approaches, all states included in this study preferred multilateral approaches to both national defence and expeditionary warfare. Only one state, Poland, gave equal priority to both approaches when it comes to national defence. Arguably, neither the 11 new allies' different experiences of lasting independence nor the various experiences regarding territorial integrity and independence during WWII can explain these similarities and differences when it comes to ways.

16.3.5 Conclusions regarding the historical approach

Regarding *alignment strategies*, we noticed a covariation between, on the one hand, historical experiences of having the own states' territories completely annexed and/or controlled by other states and, on the other hand, a preference for military-oriented approaches to multiple-courting and offensive bandwagoning. However, when it comes to relations to specific states, our cross-cases comparison indicates that WWII experiences are not as important to the choice of alignment strategy as sometimes has been expected in previous research. Potentially, this observation suggests that WWII experiences may have been overshadowed by the 11 new allies' Cold War era experiences. Moreover, membership in the EU and NATO are likely to have facilitated cooperation between former wartime enemies.

Regarding *ends*, we observe a similar pattern of covariance between states suffering annexation during WWII and priorities related to survival and territorial security. The cross-cases comparison using the overlapping approach to historical experiences that focus on experiences of independence before WWII and experiences of lasting independence during the Cold War revealed that five of the six states that lacked these experiences had survival as their prioritised end. However, Slovenia broke this pattern in both approaches to historical experiences by prioritising both survival and status. In addition, Poland also pursued both survival and status.

When it comes to *means*, the cross-cases comparison showed no covariation between experiences and priorities between national defence and expeditionary warfare. However, the comparison presents a relatively strong covariance between experiences of complete annexation and/or loss of independence during WWII on the one hand, and, on the other hand, priority to military capacities needed to fight a qualified adversary. Moreover, with Hungary as the only exception, all states that escaped total annexation and/or losing their political independence during WWII have been more prepared to develop military capabilities and capacities needed for expeditionary warfare. Consequently, when it comes to priorities

200 *Explaining the findings*

regarding nationally controlled military means, experiences from WWII seem to provide stronger correlations than experiences related to lasting independence.

Finally, when it comes to *ways*, the 11 new allies' different historical experiences cannot explain the similarities and differences regarding preferences related to the unilateral and multilateral approaches.

16.4 Summarising the influence of the three intervening variables

In Table 16.2, our findings concerning covariation between the three intervening variables and differences and similarities in strategic priorities between our 11 cases are summarised.

The *positional approach* cannot contribute to explanations regarding different priorities when it comes to choosing between military-oriented approaches and broad security-oriented approaches to multiple-courting. However, a focus on the 11 allies' relative power positions reveals that nine out of ten minor middle powers and small states pursued a multiple-courting strategy. Poland, which we have upgraded to a middle power, pursued an offensive bandwagoning strategy made possible by its greater access to military means. Regarding ends, six of the seven small states prioritised survival as their only overarching end. With the exception of Romania, which prioritised status, all minor middle powers and Poland prioritised survival and an additional end related to increased influence or status, thus indicating that ambitions grow with increases in relative power. Regarding means and ways, all minor middle powers, except Hungary, have higher ambitions when it comes to transforming their armed forces and developing the capacities needed for expeditionary warfare and to providing substantial contributions to international PSOs. None of the states classified as small states were able to present convincing ambitions and capacities in this regard. Arguably, the lack of military capacities of both small states and minor middle powers make multilateral approaches to both national defence and expeditionary warfare a necessity rather than an option. Notably, Poland, the only state that we consider had a balanced approach in this regard, is also the only state having both the quantity and the quality needed to make a complementary unilateral approach to national defence credible. However, since 2015, the Romanian government, with a substantial quantity of armed forces at its disposal, has expressed ambitions of creating unilateral capacities to deter and counter possible aggression, and it has already proved itself capable of leaving substantial contributions to international PSOs.

The *geographical approach* complements the positional approach by revealing a covariation between differences in geographical position and differences in individual states' strategic priorities that are not related to differences in relative power. In this study, we limited the use of this variable to just the one geographical aspect, that is a shared land border with Russia. Regarding alignment strategy, the three Baltic States all prioritised a military-oriented approach to multiple-courting. The fourth state with borders to Russia, Poland, prioritised its strategic

Table 16.2 Intervening variables and the outcomes of the defence strategy.

	Alignment strategies	Ends	Means	Ways
Positional approach	The choice of *multiple-courting* covariates with the limited power resources among small states and minor middle powers. The only middle power pursued *offensive bandwagoning*.	Increased ambitions regarding *status* and *influence* among minor middle powers and middle powers.	Increased relative power covariates with increased ambitions for defence transformation related to *national defence* and/or to *expeditionary warfare*.	Lack of military capacities among small states and minor middle powers creates preferences for *multilateral approaches* to both national defence and expeditionary warfare.
Geographical approach	Land borders to Russia covariates with military approaches to *multiple-courting*.	Land borders to Russia covariates with priority of *survival* related to territorial integrity.	Land borders to Russia covariates with priority of means related to *national defence*.	Land border to Russia does not correlate with differences related to *uni-* and *multilateral approaches*.
Historical approach	Historical experiences of being completely annexed and/or controlled by other states during WWII covariate with preferences for military-oriented approaches to *multiple-courting* and *offensive bandwagoning*.	Historical experiences of being completely annexed and/or controlled by other states during WWII covariate with priorities related to *survival* and territorial security. States with continued independence during the Cold War more inclined to prioritise *status* and *influence*.	Historical experiences of being completely annexed and/or controlled by other states during WWII correlates with a prioritisation of means needed to fight a qualified adversary in national defence.	Historical experiences do not covariate with preferences related to *unilateral* and *multilateral* approaches.

partnership with the US and participation in the US ballistic missile defence program. Both Poland and the Baltic States have received allied troops on their soil as part of NATO's forward presence. However, Poland's access to greater military resources has enabled the Polish government to provide larger contributions to US-led coalitions, hence pursuing an offensive bandwagoning strategy. The five states that are practicing a broader security-oriented approach to multiple-courting all enjoy a less geographically exposed position. However, two additional states prioritise a military approach to multiple-courting, Bulgaria and Romania. The alignment strategies of these two states indicate that the geographical aspect of strategic exposure potentially should include additional troublesome aspects when measuring a particular state's strategic exposure. Regarding ends, all new allies except Romania gave priority to survival. However, the three Baltic States constitute a separate group among the new allies that relate *survival* primarily to the territorial integrity of the state and to concerns related to Russian minorities. Arguably, these states' shared land borders with Russia influences both preferences. Moreover, all four states with land borders to Russia have been reluctant to dismantle capacities related to national defence, and they all responded to Russia's war against Ukraine in 2014 by increasing their efforts to develop military capabilities needed to defend themselves against a qualified adversary such as Russia. Before Russia's attack on Ukraine in 2014, six of the seven states without land borders to Russia, prioritised military means and capabilities related to expeditionary warfare. As a part of their defence transformation processes, these less geographically exposed states also dramatically reduced the size of their armed forces and weapon systems related to national defence against a qualified opponent. Notably, Hungary is an exception in this regard since the Hungarian government has been rather unwilling to allocate resources for defence transformation regardless of a focus on national defence or expeditionary warfare. Regarding ways, it is difficult to find any covariation between geographical position and the preference on unilateral or multilateral approach.

Our third intervening variable, *historical experiences*, provides an additional layer complementing the two other variables. Regarding alignment strategies, the cross-case analysis revealed a covariation between, on the one hand, historical experiences of having the states' own territory completely annexed and/or controlled by other states and, on the other hand, a preference for military-oriented approaches to both multiple-courting and offensive bandwagoning. Our analysis indicates that WWII experiences are not as consequential for the choice of alignment strategy as previous research has given reason to expect. One possible explanation to this observation is that the experiences gained during WWII may have been overshadowed by later experiences, especially those from the Cold War era. However, membership in the EU and NATO are also likely to have facilitated cooperation between former enemies. Regarding *ends* there was a similar pattern of covariance between states suffering total annexation and priorities related to survival and territorial security. The comparison that focuses on lasting independence during the Cold War and experiences of statehood before WWII revealed that five of the six states that either were annexed

Explaining the diversity of strategic responses 203

by the USSR and/or lacked the experience of being independent before the end of the Cold War, focused solely on survival as a prioritised end. Regarding means, the comparison of historical experiences revealed a relatively strong covariance between, on the one hand, experiences of complete annexation and/or loss of independence during WWII and, on the other hand, priority given to military capacities needed to counter aggression of a resourceful state adversary. Moreover, all states, except Hungary, that avoided suffering a total annexation and/or loss of independence during WWII have been more willing to develop the military capabilities and capacities needed for expeditionary warfare. Regarding ways, the different historical experiences of the 11 new allies do not covariate with unilateral and multilateral approaches.

16.5 Conclusions: the outcomes of the dependent variable

As observed in Chapter 15, the total number of different favoured defence strategies turned out to be seven. However, if just Czechia's preferences following Russia's illegal invasion of Ukraine in 2014 are taken into account and the previous Czech position neglected, the number of different defence strategies is only six. Moreover, if Hungary, despite all unsuccessful attempts at military transformation, nevertheless were to be interpreted slightly differently as discussed next, the number would decrease to only five different strategies.

In the preceding sections, we have elaborated on only one intervening variable at a time. Consequently, we have not discussed the potential influence of covariation between two or even three of these variables. In this section, we turn our attention to the six different defence strategies in order to present an initial exploration of potential covariation. Notably, two or several states share three of the defence strategies, while the remaining three strategies are individual. Moreover, all three shared strategies are similar when it comes to both alignment strategy (multiple-courting) and the ways element of the military strategy (multilateral) approach. Consequently, the only variations between the three shared defence strategies are to be found in the ends and means elements.

From 2014 and onwards, Czechia, Estonia, Latvia and Lithuania preferred similar defence strategies. They all give priority to survival regarding the ends and to national defence when it comes to the means. However, contrary to the three Baltic States, Czechia is a minor middle power without shared land borders with Russia. Consequently, the only shared intervening attribute is the historical experience of being annexed during WWII. Interaction between the intervening variables is hence, in these cases, not considered as providing any additional explanatory power regarding the preferences of defence strategy.

Bulgaria and Croatia also share defence strategy with both states prioritising survival and expeditionary warfare. Notably, they also share the outcomes of all three intervening variables, that is being small states, lacking land borders with Russia and avoiding annexation during WWII. Finally, regarding shared defence strategy, Romania and Slovakia are also to be included in this group of new allies. In their case, status is the prioritised end while expeditionary warfare is the preferred design of the military means. If we decrease the importance of hosting foreign embassies, Slovakia would be categorised similarly as Romania. If so, Romania and Slovakia would share the outcomes of all three intervening variables, that is being minor middle powers, lacking land borders with Russia and avoiding annexation during WWII. If Hungary had put deeds behind the words transforming its military towards expeditionary warfare, its defence strategy would have been similar to those of Romania and Slovakia. Notably, Hungary also shares the outcomes of the intervening variables with these two states. Consequently, these five states would share all aspects of the defence strategy except regarding the ends. Potentially, their different power resources and positions in the international system could explain this difference.

Finally, Poland and Slovenia represent the two remaining and individual defence strategies. As previously discussed in this chapter, Poland should potentially be reclassified as a middle power. This would not only leave Poland in a class of its own among the new European allies but also explain Poland's preferences. Notably, these preferences are more ambitious compared to those of the other new allies regarding both the alignment strategy and the military strategy. This leaves us with Slovenia. The passive bandwagoning, in combination with rather ambitious ends, make us conclude that this new ally is an example of freeriding or buck-passing. Having the highest GDP per capita of all new allies, Slovenia is in the unique position of doing more than what has been delivered so far. Notably, Slovenia is, together with the three Baltic States, the new allies with the lowest military expenditure. All these four new allies have historical experiences of being annexed during WWII. Moreover, these states also have similar weak positions in the international system. However, the Slovenian government, in contrast to its colleagues in the Baltic States, does not perceive its country as being strategically exposed. We conclude that Slovenia's favourable geographical position, being embedded among EU and/or NATO allies, most likely contributes to these perceptions.

Note

1 Prominent examples of research on small states focusing on differences in relative power include Baker Fox (1959), Vital (1967) and Rothstein (1968). For a survey on later and different approaches to small state research, see Neumann and Gstöhl (2006) and Edström *et al.* (2019). For a classical study in middle powers, see Holbraad (1984). Gilley and O'Neil (2014) and Edström and Westberg (2020) provide examples of more recent research using a positional approach.

Bibliography

Baker Fox, Annette (1959). *The Power of Small States: Diplomacy in World War II*. Chicago: University of Chicago Press.

Brezhnev, Leonid (2006). 'The Brezhnev Doctrine' in Gearóid Tuathail, Simon Dalby and Paul Routledge (eds). *The Geopolitics Reader*. Abingdon: Routledge.

Bush, George (1991). *State of the Union.* Transcript of the President's Message to the Nation 29 January 1991, published by The New York Times 30 January 1991.

Edström, Håkan, Dennis Gyllensporre and Jacob Westberg (2019). *Military Strategy of Small States: Responding to the External Shocks of the 21st Century.* Abingdon: Routledge.

Edström, Håkan and Jacob Westberg (2020). *Military Strategy of Middle Powers: Competing for Security, Influence and Status in the 21st Century.* Abingdon: Routledge.

Gilley, Bruce and Andrew O'Neil (eds) (2014). *Middle Powers and the Rise of China.* Washington, DC: Georgetown University Press.

Gray, Colin (2006). *Strategy and History: Essays on Theory and Practice.* Abingdon: Routledge.

―――― (2015). *The Future of Strategy.* Cambridge: Polity Press.

Holbraad, Carsten (1984) *Middle Powers in International Politics.* London: Macmillan Press.

International Institute for Strategic Studies (IISS) (2020). *The Military Balance 2020.* London: Routledge.

Kier, Elizabeth (1997). *Imagining War: French and British Military Doctrine between the Wars.* Princeton, NJ: Princeton University Press.

Lašas, Ainius. (2010). *European Union and NATO Expansion: Central and Eastern Europe.* New York: Palgrave Macmillan.

Layne, Christopher (2012). 'This Time It's Real: The End of Unipolarity and the "Pax Americana"' *International Studies Quarterly* Volume 56, Issue 1.

Levy, Jack (1994). 'Learning and Foreign Policy: Sweeping a Conceptual Minefield' *International Organization* Volume 48, Issue 2.

Mearsheimer, John (2018). *The Great Delusion: Liberal Dreams and International Realities.* New Haven, CT: Yale University Press.

Morgenthau, Hans (2006). *Politics among Nations: The Struggle for Power and Peace.* New York: McGraw-Hill.

Neumann, Iver and Sieglinde Gstöhl (2006). 'Lilliputians in Gulliver's World?' in Christine Ingebritsen, Iver Neumann, Sieglinde Gstöhl and Jessica Beyer (eds). *Small States in International Relations.* Reykjavik: University of Iceland Press.

Reiter, Dan (1994). 'Learning, Realism, and Alliances: The Weight of the Shadow of the Past' *World Politics* Volume 46, Issue 4.

Rothstein, Robert (1968). *Alliances and Small Powers.* New York: Columbia University Press.

Snyder, Jack (1991). *Myths of Empire: Domestic Politics and International Ambition.* Ithaca, NY: Cornell University Press.

Truman, Harry (2006). 'The Truman Doctrine' in Gearóid Tuathail, Simon Dalby and Paul Routledge (eds). *The Geopolitics Reader.* Abingdon: Routledge.

Tuathail, Gearóid (2006). 'General Introduction: Thinking Critical about Geopolitics' in Gearóid Tuathail, Simon Dalby and Paul Routledge (eds). *The Geopolitics Reader.* Abingdon: Routledge.

Vital, David (1967). *The Inequality of States: A Study of the Small Power in International Relations.* Oxford: Calderon Press.

17 Conclusions

Strategic responses to membership demands and changes in the external security environment

Having discussed and analysed the influence of our three intervening variables in the previous chapter, we now conclude our exploration by addressing the main question presented in the introduction:

> How have the 11 new allies responded to the systematic pressures emanating from their membership in the two alliances and the changes in their external security environment during the two first decades of the twenty-first century?

The question is answered in two sections. The first of these sections focuses on the 11 new allies' responses to the systematic pressures related to the 'new European security dynamics' and the membership processes to develop and improve their capability to contribute to allied efforts related to international peace support operations (PSOs) and expeditionary warfare in general. In addition, in this section we also reflect on our methodological considerations presented in Chapter 1. The second of these sections focuses on responses to the deteriorating European security environment that have created a renewed need for military capabilities related to national and collective defence against a qualified state opponent. In the third and final section, we present our aggregated conclusions.

17.1 The new European security dynamics and membership processes

For many contemporary observers, the Yugoslav Wars of independence 1991–2001 was assumed to be the last chapter in the long book on reoccurring European wars. In a study on 'the new European security dynamics' from 2009, Janne Haaland Matlary argued that the 'nation-state model of defence' had been replaced with a new 'post-national' ideas regarding of the use of armed force. Previous conceptions of war as existential struggles between competing nation-states was replaced by a new kind of 'optional wars' where states choose to participate in and contribute to international operations far away from their own borders rather than defending their own territory and citizens. Consequently, the notion of national territorial defence gradually became less relevant for the actual use of military force (Matlary 2009:3 and 24–25; see also Diesen 2005; Smith 2007).

According to Timothy Edmunds, the early post–Cold War era was characterised by a 'significant shift' towards new kinds of missions for the armed forces of the European states. These new missions aimed to counter threats related to international terrorism, regional instability caused by intra-state conflicts, the supply of strategic resources and the proliferation of weapons of mass destruction (WMD). To be able to contribute to these multilateral military operations, states were forced to develop capacities related to expeditionary warfare, including flexible and technologically advanced force structures with professional soldiers able to interact with the armed forces of other participating states. Institutionally, pressures to develop these capabilities were channelized programs such as NATO's Partnership for Peace (PfP) and Membership Action Plan (MAP). Within the EU, the 1999 Helsinki Head Line Goals provided additional impetus to these defence transformation processes (Edmunds 2005:10–11 and 14). The defence planning of the Warsaw Pact (WP) had, according to Jeffrey Simon, left many of the new allies with decaying Soviet technology and an oversized force structure that was incompatible to NATO's new operational needs (Simon 2001:29–31). The membership criteria of the EU partly overlap with those of NATO regarding well functioning democratic systems, the rule of law and market economy. However, in contrast to NATO, the membership criteria of the EU do not concern military capabilities or civil–military relations (European Council 1993).

As elaborated in Chapter 2, previous research on European integration processes have identified a number of different mechanisms related to how common institutions affect the policies of member states. Kyriakos Moumoutzis and Sotirios Zartaloudis distinguished between two distinct types of learning processes: 'instrumental learning' that concerns new and more cost-effective use of means and 'social learning' that concerns a change in policy ends and objectives (Moumoutzis and Zartaloudis 2016). Socialisation has been presented as an alternative concept to social learning. According to Nicole Alecu de Flers and Patrick Müller, socialisation can be understood as a process whereby members of a community are 'inducted into the community's rules, norms and policy paradigms' (de Flers and Müller 2012:24). If this distinction is applied to defence strategies, we expect that the strategies of the new allies can be influenced by their membership processes in two different ways. First, they may internalise the liberal Western objectives and values common to both organisations. Second, the new allies may adjust their strategies to reach a more common cost-effective use of means and specific policy paradigms such as NATO's Comprehensive Operations Planning Directive (COPD) or the EU Battle Group concept. Membership in NATO also creates opportunities for 'instrumental learning', when it comes to questions related to collective defence, such as access to commonly owned resources as well as burden sharing with certain states offering specific niche capacities, relying on other states to provide support with other capacities. In this study, we focus on the aspects of learning and socialisation related to the strategic ends.

The new security dynamics have potentially far-reaching effects on individual states' alignment and military strategies. Regarding *alignment strategies*, membership in both organisations, offers opportunities to share risk, cost and responsibility in relation to participation in international PSOs (Matlary 2009). Small states and minor middle powers that have internalised worldviews and ends associated with the post-national security are expected to pursue multiple-courting strategies along the lines of the broader security-oriented approach discussed in Chapter 16. Regarding strategic *ends*, the new security dynamics and pressures related to the membership processes during the first decade of the twenty-first century, suggest that states should give higher-priority goals related to influence and status. Moreover, the priorities of the new allies may also reflect an internalisation of the liberal values common to the EU and NATO. When it comes to *means*, the new dynamics suggest that even less resourceful states will increase their efforts to develop capacities for expeditionary warfare, making them cable of contributing to PSOs or coalitions of the willing led by more resourceful states. As elaborated on in Chapter 3, the means related to expeditionary warfare include interoperability with allied forces, high skill levels of the soldiers as well as access to logistic chains and transport systems for deployment of lighter military units outside the states' own regions. To afford this defence transformation, most states had to reduce the size of their armed forces and limit the amount of heavy military equipment associated with twentieth-century warfare between nation-states. Regarding *ways*, the new European security dynamics suggest a decreased emphasis on unilateral approaches and an increased emphasis on multilateral approaches. However, as will be further discussed in the second section of this chapter, governments may disagree with the assumptions of a new stable and peaceful European security order and continue to perceive that their states are exposed to threats from other states. This category of states is less likely to follow the path of the new security dynamics and reduce the capabilities needed for traditional interstate warfare. Other states may find that their own lack of resources prevent them from developing military capacities related to expeditionary warfare.

In their analysis of the strategic environment during the first two decades of the twenty-first century, all new allies except Estonia predominately expressed concerns primarily related to non-state actors such as international terrorism, organised crime, proliferation of WMD, trafficking and illegal migration. The sources of these non-state and transnational threats were often located in unstable regions and in authoritarian or failed states, creating incitements and legitimacy for international military operations. This broadened conceptualisation of security also included non-antagonistic threats such as the spread of epidemics, degradation of the natural environment and climate change. Before Russia's war against Georgia in 2008, a clear majority of the new allies estimated that a large-scale military attack directed against any NATO member was highly unlikely within a predicable time horizon. In 2007, the Polish government expressed concerns about Russia's attempt to reinforce its position. However, the government still concluded that a large-scale armed conflict was unlikely in the foreseeable future. In 2005, the Latvian government concluded that there was no direct military threat to either Latvia itself or to any other Baltic State. In 2006, the government of Lithuania reported that new unconventional security challenges were replacing traditional threats of armed aggression against the territory of sovereign states. Estonia, the only new ally that clearly went against this trend of embracing the threat conceptions of the new European security paradigm, already in 2004 warned that the new international security environment had not reached a state of stability. Consequently, the Estonian government concluded that it had to prepare for possible military threats related to the unexpected redeployment of military forces and large-scale military manoeuvres near Estonian territory as well as intentional violations of Estonia's air space, land border and territorial waters.

To which extent were these changing threat perceptions followed by changes in the defence strategies of the new allies? As previously mentioned, all the new allies, except Poland and Slovenia, pursued the alignment strategy of multiple-courting and actively cooperating with both the EU and NATO. However, as discussed in Chapter 16, states with a more geographically exposed position preferred a military-oriented approach to multiple-courting focusing primarily on NATO and bilateral defence cooperation. Apparently, even though Latvia, Lithuania and Poland did not express concerns regarding military aggression against themselves or other members of the EU and NATO before 2008, they still hesitated to embrace the broader security-oriented approach pursued by the states that had a less geographically exposed position.

Regarding strategic *ends*, more states than expected continued to prioritise survival. As previously discussed, more ambitious aims related to status and influence were pursued primarily by the states classified as minor middle powers or middle powers. Consequently, our results indicate that the 'post-national' security paradigm was not fully internalised among the new allies. Moreover, while the three Baltic States primarily related survival to their territorial integrity, less geographically exposed states instead related survival to national identity and diasporas living in neighbouring states. Regarding the internalisation of liberal values such as democracy, human rights, rule of law and free market economy, all new allies expressed support for these principles both before and after joining the EU and NATO. However, the priority of these values compared to other ends such as national identity and traditions were slightly different. Some states, such as the Baltic States, Bulgaria, Czechia and Slovakia, gave these values the status of vital interest, while other states, such as Croatia, Hungary, Poland and Slovenia, gave even greater priority to objectives related to sovereignty, national integrity and identity.

When it comes to the *means* and the responses by the new allies to pressures regarding transforming their armed forces in order to be able to contribute to international PSOs, some elements are common to most states. All states present, for example, ambitions to contribute to international PSOs and have at least part of their armed forces manned with professional soldiers. With the exception of the three Baltic States, Czechia and Slovakia, all the new allies have made significant reductions in the size of their land forces. Czechia, which decreased the number of brigades from three in 2000 to two in 2020, seems to have undertaken the reduction of the size of the Czech army already during the 1990s. Contrary to Czechia, both Estonia and Lithuania have increased the size of their army during

210 *Explaining the findings*

the last decade, while both Latvia and Slovakia have kept the same levels since the beginning of the new millennium. In addition, as presented in Table 17.1, we observe significant reductions in numbers related to some key equipment such as main battle tanks (MTBs) and combat aircraft.

The downsizing of land and air forces were in most cases motivated by the need to modernise the armed forces and/or to enable the allocation of resources for developing new capabilities. In some cases, for example Romania and Slovakia, there were also explicit references to EU and NATO standards and membership criteria. When it comes to means, a majority of the new allies adjusted their force structure to expectations related to the membership processes as well as to the assumed stable and peaceful European security order. However, as mentioned in previous chapters, ambitions and actual ability to contribute to common efforts within the EU and NATO vary among the new allies. The comparably more resourceful states, that is Czechia, Poland and Romania, have generally been more willing and capable of delivering what has been pledged, while other states, such as Hungary, Slovakia and Slovenia, have openly admitted failure to allocate the resources needed for their defence transformation processes. Arguably, both Bulgaria and Croatia have limited their ambitions after becoming members. Moreover, several states have continued to rely on decaying USSR equipment.

Regarding *ways*, the new European security dynamics and the post-national security paradigm emphasise the need for multilateral cooperation when conducting international operations. For small states and minor middle powers, unilateral approaches are not an alternative when it comes to expeditionary warfare and even the most resourceful state among the new allies, Poland, has not expressed any unilateral ambitions related to international operations. However, Poland has, on several occasions, assumed the role as framework nation for EU battle groups.

So what does this mean for military strategy at an aggregated level? Since most of the states explored have rather weak economies, budget deficits may very well have forced the governments to fiscal austerities including spending cuts. In addition, other security challenges such as migration crises can potentially have led to other priorities within the overarching security policy. Moreover, the governments

Table 17.1 Number of MBTs and combat aircraft of some of the new allies.

	Main battle tanks		Combat aircraft	
	2000	2020	2000	2020
Bulgaria	1,475	90	181	21
Croatia	300	75	41	11
Czechia	792	30	110	38
Hungary	806	44	68	14
Poland	1,704	606	267	95
Romania	1,253	400	323	56
Slovakia	275	30	84	23

Source: International Institute for Strategic Studies (2000, 2020).

also have had to deal with the paradox of strategy, that is balancing the current needs of using military force, on the one hand, with the needs of generating military force for addressing future threats, on the other hand. Perceptions of immediate security risks may have, for example, been met with rapid military operations and responses. In order to finance these activities, funding for the intended procurements of the armed forces may have been reallocated. Consequently, several different circumstances can have forced the governments to adjust the declared ambitions in the primary sources explored.

As previously mentioned, such adjustments seem to have been undertaken in some of the cases explored. Arguably, Bulgaria, Croatia, Hungary, Slovenia and Slovakia have not fully allocated the funding needed for implementing the ambitions articulated in their official documents. When comparing the figures of the governments with those of independent sources, such as the IISS, the means at the disposal of the armed forces tend to be below the intended levels in each of these countries. In the three latter cases, the government has explicitly admitted its shortcomings in this regard.

Notably, all these five new allies have rather favourable geographical locations with respect to a potential Russian threat. Potentially, this intervening variable may have some explanatory power in this regard.

Another interesting aspect is connected to the balancing of the three elements of the military strategy. Arguably, the three Baltic States as well as Czechia, Poland and Romania have indicated not only a mature strategic awareness but also nuanced elaborations regarding the potential need for adjusting the ends, means and/or ways depending on changes in the strategic environment during the period explored. This has not always been the case when it comes to Bulgaria, Croatia, Hungary, Slovenia and Slovakia. Consequently, despite the admitted shortcomings, allocating the funding necessary to transform the means of the armed forces, internal efforts has not been undertaken to revise the ends and/or ways in order to re-establish a balance between the three elements. Instead, the balancing act seems to have been undertaken between the internal and external efforts. Potentially, the alignment strategy may thus include aspects of free-riding due to a lack of perceived strategic exposure.

17.2 Responses to a deteriorating European security order

Arguably, the reluctance of some of the new allies to adjust to the new security dynamics and to undertake defence transformation towards expeditionary warfare are related to changes in the regional pattern of interactions and war expectations from 2008 and onwards. In our study on the Nordic countries' responses to Russia's military aggression in 2008 and 2014, we observed an increased emphasis on ends and means related to national and collective defence. We also notified intensified external efforts for defence cooperation with other states and organisations during the second decade of the twenty-first century (Edström *et al*. 2019). To which extent did the new allies react to Russia's military aggression against Georgia and Ukraine?

Notably, Estonia was the only country that responded to Russia's attack against Georgia in 2008 by immediately increasing the size of its armed forces. In a report on the long-term development of the Estonian armed forces, the Ministry of Defence (MoD) emphasised that it was necessary to have the capacity to resist potential aggression until the arrival of allied forces. Moreover, the government expressed a need for increasing the capacities related to advanced air defence equipment as well as anti-tank and mechanised capabilities within the brigade's framework. Following Russia's war against Ukraine in 2014, the Estonian MoD announced several additional measures. The two other Baltic States did not react as promptly to the Russian–Georgian War. In 2012, the Latvian government declared that its priority was related to the quality of capabilities, not their size. The government also continued to elaborate on contributions to NATO-led operations and participation in EU battle groups. However, following Russia's annexation of Crimea, access to efficient intelligence, air defence, anti-tank capabilities and tactical mobility were emphasised. In 2012, the Lithuanian government still had the ambition having 50 percent of the land forces prepared to contribute to missions outside its territory. After 2014, the Lithuanian government intensified its external efforts related to collective defence by organizing regular joint exercises in order to enhance US presence in the region, by developing already existing defence relations with Poland and by promoting the establishment of a Lithuanian, Polish and Ukrainian brigade. One year later, the government announced a reformation of the Lithuanian land force.

Czechia, Poland and Romania did not respond to Russia's war against Georgia, but they all responded firmly to Russia's 2014 attack on Ukraine and annexation of Crimea. In 2015, the Czech government admitted that the personnel strength of the armed forces temporarily had decreased during the last couple of years despite the fact that the tasks had remained unchanged. Two years later, the government announced its ambitions to foster close defence cooperation with Germany. This effort included establishing multinational military units focusing on collective defence operations. Hence, Czechia intended to assign a brigade to a German army division to be organised. Moreover, the government concluded that it was essential to develop new army units and to increase the personnel strength by an additional 5,000 military professionals. In 2017, the Polish government proposed increasing the numbers of operational forces, enhanced artillery capabilities, a new generation MBTs and new air defence system. Moreover, the number of special forces and coastal missile units as well as of submarines was to increase. Regarding the air force, the Polish government prioritised new attack helicopters and new, fifth-generation combat aircraft equipped with long-range precision weapons. Within the two following years, these new priorities resulted in the acquisition of the Patriot missile defence system and 32 F-35 combat aircraft. In 2015, the Romanian government presented priorities regarding military capacities related to armed conflicts between advanced adversaries including new combat aircraft and naval vessels. Moreover, in 2016 the government announced that the new organisation of the army was to include two division headquarters. In 2020, the government presented its intentions of enabling and strengthening the NATO-enhanced forward presence on Romanian soil.

Regarding the five remaining new European allies – Bulgaria, Croatia, Hungary, Slovakia and Slovenia – it is difficult to find clear examples of firm responses either to the Russia–Georgian War or to Russia's war against Ukraine in 2014. To explain the lack of responses from this group of the new allies, we will, in the next concluding section, return to our previous discussion on intervening variables. However, this time we consider the interaction of two or more of these variables.

17.3 Strategic exposure and the future challenges facing the new European allies

In Chapter 16, we demonstrated a covariation between states with a *geographically* exposed position – operationalized as states sharing a land border with Russia – and priorities regarding military-oriented approaches to alignment strategies. All states with land borders to Russia prioritised military-oriented approaches, the Baltic States applying multiple-courting and Poland an offensive bandwagon strategy towards the US. This geographically exposed position also covariated with strategic ends prioritising territorial integrity and military means related to national or collective defence against a qualified opponent. Notably, the five new allies that did not respond firmly to either the Russia–Georgian War in 2008 or Russia's war against Ukraine in 2014 all lack a land border to Russia. Moreover, all new allies that share land borders with Russia responded especially to Russia's renewed military aggression in 2014. Land borders to Russia therefore seem to be a *sufficient* condition for provoking responses. In addition, geographical position seems to have rather strong explanatory power regarding the general direction of the new allies' responses to increased war expectations in their regional system. However, land borders with Russia is not a *necessary* condition for firm responses. Both Czechia and Rumania responded to Russia's annexation of Crimea in spite of a less geographically exposed position vis-à-vis Russia. In the case of Rumania, its land borders to Moldavia and its coastline to Black Sea may provide other reasons for the government to perceive geographical exposure. On the other hand, the Romanian government's responses may also be motivated by other concerns related to its self-image of being a trustworthy and capable EU/NATO ally. Notably, Romania is the only state for which we concluded that *status* was the prioritised end. In the case of Czechia, its historical experiences of being annexed may have contributed to the Czech government's firm response to Russia's war against Ukraine.

In the analysis of covariation between *historical experiences* and strategic priorities, we found that the experiences of being completely annexed and/or controlled by other states during WWII covariate with preferences similar to those of the states that share a land border to Russia. Regarding the Baltic States and Poland, it is not possible to isolate these two intervening variables from each other since they both share a land border to Russia and were annexed by both Germany and the USSR during WWII. For these four states, incitements related to their geographical position is likely to be reinforced by their historical experiences during both WWII and the Cold War era. Bulgaria, Croatia, Hungary and Slovakia did not suffer this fate even though they also suffered from occupation and, in some cases, had to except puppet regimes loyal to Nazi-Germany. Similar to the four states that share a land border with Russia, Czechia suffered, as previously touched upon, annexation and external political control. However, the historical experiences during WWII do not help in explaining why Czechia, following Russia's war against Ukraine in 2014, choose to integrate part of its armed forces with the state that annexed its territory during WWII, that is Germany. However, Germany, assisted by its membership in both the EU and NATO, seems to have managed to establish a break with its past dark history, which the USSR/Russia most likely has been incapable of. In this respect, our results indicate that the WWII experiences are not as important to the strategic priorities regarding alignment strategies as sometimes expected. A possible explanation to this observation is that the legacy of the WWII for many of the 11 new allies has been overshadowed by their experiences during the Cold War era, such as, in the case of Czechia, the armed intervention by the USSR and other WP countries in 1968. However, Hungary, which also suffered armed Soviet intervention in 1956, has, for example, not responded as firmly as Czechia in this regard.

Our third intervening variable, the *positional approach* focusing on differences related to relative power, is the only variable that can explain differences among the 11 new European allies' priorities related to the ways and the choice between unilateral and multilateral strategies. Due to a lack of resources, all small states and middle powers included in our selection of cases were dependent on multilateral strategies for both collective territorial defence and contributions to expeditionary warfare. Only one state, the middle power, Poland, presented balanced ambitions regarding unilateral and multilateral strategies when it came to national defence. None of the new allies contemplated unilateral ambitions when it comes to expeditionary warfare. Differences in relative power were also reflected in increased ambitions to pursue complementary aims to survive regarding strategic ends. Most importantly, differences in relative power correlated strongly with ambitions when it comes to internal efforts relating to both expeditionary warfare and territorial defence. Differences in relative power may be important in one additional way: states that are geographically exposed and that, for historical reasons, perceive themselves as threatened by a specific state are likely to view themselves as even more exposed without unilateral capacities to create cost for a potential aggressor. Russia's full-scale invasion of Ukraine on 24 February 2022 has created a situation in which several of our cases most likely will have to consider questions related to geographical exposure. Hungary, Poland, Romania and Slovakia all share, for example, borders with Ukraine. Arguably, the European states that previously followed the post-national path are ill prepared for the return of the old European security order. Considering the time gap between deciding strategies for force generation and, once these strategies are implemented, establishing strategies for the use of the new military force, these states will now face a situation where they will have to rely on external efforts and the possibility of sharing already existing military resources with like-minded countries.

Bibliography

de Flers, Nicole Alecu and Patrick Müller (2012). 'Dimensions and Mechanisms of the Europeanization of Member State Foreign Policy: State of the Art and New Research Avenues' *Journal of European Integration* Volume 34, Issue 1.

Diesen, Sverre (2005). 'Mot et alliansintegrert forsvar', in Oyvind Osterud and Janne Haaland Matlary (eds). *Mot et avnasjonalisert forsvar?* Oslo: Abstrakt forlag.

Edmunds, Timothy (2005). 'A New European Security Environment? The Evolution of Military Roles in Post-Cold War Europe', in Timothy Edmunds and Marjan Malesic (eds). *Defence Transformation in Europe: Evolving Military Roles*. Amsterdam: IOS Press.

Edström, Håkan, Dennis Gyllensporre and Jacob Westberg (2019). *Military Strategy of Small States: Responding to the External Shocks of the 21st Century*. Abingdon: Routledge.

European Council (1993). *Conclusion of the Presidency*. European Council in Copenhagen 21–22 June. (www.consilium.europa.eu/media/21225/72921.pdf).

International Institute for Strategic Studies (2000). *The Military Balance 2000–2001*. Oxford: Oxford University Press.

International Institute for Strategic Studies (2020). *The Military Balance 2020*. London: Routledge.

Matlary, Janne Haaland (2009). *European Union Security Dynamics: In the New National Interest*. Basingstoke: Palgrave Macmillan.

Moumoutzis, Kyriakos and Sotirios Zartaloudis (2016) 'Europeanization Mechanisms and Process Tracing: A Template for Empirical Research' *Journal of Common Market Studies* Volume 54, Issue 2.

Simon, Jeffrey (2001). 'NATO's Membership Action Plan and Defense Planning' *Problems of Post-Communism* May/June.

Smith, Rupert (2007). *The Utility of Force: The Art of Warfare in the Modern World*. New York: Alfred Knopf.

Index

Note: Page numbers in *italics* indicate a figure and page numbers in **bold** indicate a table on the corresponding page. Page numbers followed by 'n' indicate a note.

Afghanistan 57, 76, 99, 112, 143, 168, 177; EU Police Mission in 110; NATO-led operation in 112, 119, 134, 143; Romania's deployment to 143, 146, 178, 185
air defence systems 45, 131, 191, 212
air surveillance radars 87, 108
Alecu de Flers, Nicole 207
alignment strategies **41**, 175–177; balanced approach towards 186; balance of power 135; basic options 40; concept of 39; differences in relative power 182–188; historical approach to 195–197; military-oriented approach 176, 213; multiple-courting 180; positional approach to 182–188, 187–188; Slovakian 157; Slovenian 177; strategic priorities regarding the choice of 182; *see also* defence strategies; military strategies
Ämari Air Base, reconstruction of 85
American AEGIS Ashore system 143
An-32 transport aircraft 65
anti-ballistic defence 154
anti–sea mines operations 55
anti-surface and anti-mine warfare 65
anti-tank systems 87
Area of Operations 66, 109
armed aggression, WWII experiences of **196–197**
armed forces: competences and recruitment processes of 18; interoperability with forces from other states 45; modernisation of 9, 51; 'post-national' ideas regarding of the use of 15
armoured combat vehicles (ACVs) 55
armoured personnel carriers (APCs) 86

arms and technologies, illegal proliferation of 52
arms race 105
Austrian Empire 71
Austro-Hungarian Compromise of 1867 93
Austro-Hungarian Empire 161
Axis alliance 93
Axis powers 138, 161

balance of power 7, 28, 39–41, 43, 59, 102, 135
balance of threat theory 40–41
Balkan War (1912–1913) 51, 67
ballistic missile defence programme 134, 176, 202
Baltic States 5, 32, 111, 176, 185, 202; air policing mission 143; geographical approach, to defence planning 189; military cooperation with Nordic cooperation 112; multilateral strategies for national defence 192; multiple-courting strategy 213; NATO membership 122; potential need for adjusting the ends 211; priority of survival and territorial integrity 193; survival related to territorial security 197; territorial integrity of 209; trilateral cooperation among themselves 189
bandwagoning strategy: appeasement policy 40; Croatia 67; defensive 40; offensive 40, 176, 183, 189, 195, 200, 202; passive 168, 177, 184, 195; Polish 135; for profit 40; US-oriented 176
Baylis, John 41
Beaufre, André 41
Beeres, Robert 122

Bell, Joseph 101, 135
BENELUX countries 15
Bennet, Andrew 11n8
Berlin, Treaty of (1878) 5, 51
bilateral defence cooperation 89, 112, 209
bipolar world order 94
Black Sea Naval Cooperation Group (BLACKSEAFOR) 142
black trinity 23, 188
Borisov, Boyko 55
Bosnia and Herzegovina 64, 68, 99, 143
Brezhnev doctrine 23, 188
broader security-oriented approach 176, 183, 189, 193–194, 202, 208–209
Brussels, Treaty of (1965) 3
Bucharest, Treaty of (1913) 51
buck-passing strategy 40
Bulgarian Empire 51
Bulgaria, strategy of 51–59, **58**; defence white papers (DWPs) 51, 52; fundamental ends of 53–54; historical background 51; main military resources 56; means of 54–56; memberships in NATO/EU 53, 58; modernization of armed forces 51; multiple-courting 57; 'NATO and EU First' approach to defence policy 55; operations within the territory of other NATO member states 57; strategic environment 52–53; ways of 56–57
Bush, George H. W.: conceptualisations of a new US-led liberal world 188; vision of 'Europe whole and free' 23, 188
Byzantine Empire 161

Celestine III, Pope 82
Central Europe 5, 23, 63, 71, 129, 152
chain gang 40
Chappell, Laura 134–135
chemical, biological, radiological and nuclear (CRBN) defence 65
chemical protection battalion 54
Christian orders 104, 115
Churchill, Winston 3
civil–military relations 19, 207
civil war 8, 16, 32, 117
Coast Guard Ship Flotilla 108
Cold War 3, 106, 151, 188; events leading to the end of 11n3; military conflict during 59; NATO membership in the post–Cold War era 18; Nordic countries' defence strategies during 31; power competition between US and USSR 14; security order 21
collective security 7, 21, 28, 84, 167–168

collective self-defence 4, 45; efforts related to 8, 33; NATO's strategy for 46n1, 180; support and preparations for 30
combat engineer battalion 54
common defence planning 4
Common Foreign and Security Policy (CFSP) 99
Commonwealth of Independent States (CIS) 94, 107
constitutional order, preservation of 96
Copenhagen criteria, for membership of EU 19
Council of Europe 99
counter-insurgency, strategy for 45
coup d'état 82, 104, 115
courting, strategy of 40
COVID-19 pandemic 140, 152
Crimea, Russia's annexation of 4, 78, 117, 140, 177, 212
crisis management operations 87
Croatian armed forces, development of 61
Croatian crown 61
Croatian War of Independence 61
Croatia, strategy of 61–69, 198; armed forces, role of 66; bandwagoning 67; building offensive naval capabilities 69; chemical, biological, radiological and nuclear (CRBN) defence 65; competition with Venice for control over the eastern Adriatic coast 61; defence budget 68; Erdut Agreement (1995) 61; expeditionary capabilities 68; foreign policy 66; fundamental ends of 63–64; historical background 61; Hungarian invasion 61; main military resources **66**; means of 64–66; Membership Action Plan (MAP) 69; memberships in NATO/EU 67; military capabilities 64; military strategy risks 69; multilateral approach 68; multiple-courting 67; national cyber security strategy (NCSS) 61; Ottoman invasions 61; program of international defence co-operation 66; for protection of Croats living outside the borders of Croatia 64; on relation with US 69; seat in UNSC 69; security system 64; strategic environment 62–63; submarine warfare capability 65; War of the Croatian Succession 61; ways of 66–67
cultural heritage 58, 96, 118, 141, 177, 184
cyber-attacks 83, 140
cybercrime 53
cybersecurity 95
cyberspace 62, 163

cyberterrorism 63, 105
Czech and Slovak KFOR (Kosovo Force) battalion 76
Czechoslovakia 150, 177, 184, 195, 198; *ad hoc* coalition groupings 76; Armed Forces of 73; cyber defence of 75; dissolution of 72; goals of military reforms 79; Gross Domestic Products (GDP) 72; main military resources of **76**; response to Russia's war against Ukraine 213; Soviet occupation of 71; strategic interests 73; Velvet Revolution (1989) 71–72; war with Poland 126
Czech Republic (Czechia), strategy of 71–79; for crisis management operations 75; difference with Hungary's defence strategies 188; for economic and social development 73; for ensuring the security of citizens 74; fundamental ends of 73–74; historical background 71–72; means of 74–76; multiple-courting alignment strategy 77; for protection and defence of airspace 76; against risks of cyber attacks 73; Saint-Germain-en-Laye, Treaty of (1919) 71; strategic environment 72–73; for strengthening of deployable capabilities 75; uprising against the communist regime 71; ways of 76–77

Dănilă, Ştefan 147
Danish Duchy of Estonia 82
de Bakker, Eric 122
decision-making 28, 43, **88**
defeat in war 30
defence strategies: definition of 10, 39; intervening variables and the outcomes of **201**; of Nordic countries during Cold War 31; *see also* alignment strategies; military strategies
defence transformation 17, 202; in response to a deteriorating European security order 211–213; towards expeditionary warfare 211
defence white papers (DWPs) 9, 51, 118
defensive realism, characteristic of 42
defensive realists 42
de Flers, Nicole Alecu 20
democratic statehood, strengthening of 107
Diesen, Sverre 16
"dividing lines" in Europe, idea of 18
division of labour 24
Doeser, Fredrik 135

domestic security 105–106
Dotzev, Nikolay 58
drugs trafficking 72
Duchy of Bohemia 71

Eastern Europe 3, 5, 19, 52–53, 73, 95, 142, 162, 175
Eastern military bloc 14; *see also* Warsaw Pact (WP)
East–West division, of Europe 15
economic dimension of security 177
economic power resources 26
economic security 74, 105, 117, 147, 154
Edmunds, Timothy 17–18, 207
Eidenfalk, Joakim 135
ends: Baltic States political need to adjust 211; Bulgarian strategy 53–54; Croatian strategy 63–64; Czech Republic (Czechia), strategy of 73–74; element of the military strategy 177–178; Estonian strategy 84–85; geographical approach 189; historical approach 197–198; Hungarian strategy 95–96; Latvian strategy 106–107; Lithuanian strategy 117–118; military strategies 177–178; Polish strategy 128–129; positional approach 184; Romanian strategy 140–142; Romania's priorities regarding 189; Slovakian strategy 152–153; Slovenian strategy 163–164; strategic 184
energy sources, diversification of 53
Enhanced Forward Presence 75, 144, 192
Erdut Agreement (1995) 61
Estonia, strategy of 82–90, **89**, 209, 212; against cyber attack 83; defence development plans 82; fundamental ends of 84–85; historical background 82; Livonian Order 82; main military resources **86**; means of 85–87; membership in NATO and the EU 87; for military defence 86; multiple-courting 88; for national defence 89; pillars of military defence 87; response to Russia's attack against Georgia 212; strategic environment 82–84; war of independence against the Soviets 82; ways of 87–88
ethnic conflicts 104–105
EU Battle Groups (EUBG)s 56, 64–65, 74, 76, 88, 143; concept of 207; Hungarian contributions to 98; Lithuanian contributions to 119
Eurasian 'pivot area' 22

Euratom *see* European Atomic Energy Community (EAEC)
Euro-Atlantic area 106, 118, 152, 162
Euro-Atlantic collective security system 84
Euro-Atlantic integration 53, 66, 94, 98–99, 118
Euro-Atlantic security 147
European Atomic Energy Community (EAEC) 3
European Coal and Steel Community (ECSC) 3, 15
European Communities (EC) 3, 15
European Council 19
European Economic Community (EEC) 3, 15
European great powers 15
European integration processes 207
Europeanisation, processes of 19–20
European Rapid Reaction Force 133
European Security and Defence Policy (ESDP) 73, 133
European security dynamics 45, 188, 210; and membership processes 206–211; new and old 14–22; new post-national security dynamics 15–21; return of 21–22
European security order 4, 208; defence transformation towards expeditionary warfare 211; responses to a deteriorating 211–213
European Union (EU) 14, 52, 99; Bulgaria's accession to 53, 58; Common Security and Defence Policy of 57, 88; Copenhagen criteria for membership of 19; crisis management capabilities 77; Croatia's accession to 67; Estonia's accession to 87; foreign policy *acquis* 20; Germany's membership in 197; Helsinki Head Line Goals (1999) 18, 207; Latvia's membership in 105; Lithuanian membership of 118; membership of 9–10, 207; Member States 57; Military Rapid Response Concept 56; Monitoring Mission in Georgia (EUMM Georgia) 110; operation in Macedonia 110; Permanent Structured Cooperation (PESCO) 56; Police Mission in Afghanistan (EUPOL Afghanistan) 110; policy in relation to Russia 41; Polish membership of 127; Pooling and Sharing initiatives 77; Romania's membership of 140, 146; Slovenia's membership of 163, 169
European Union Force (EUFOR) 99

existential wars, idea of 16
expeditionary force, development of 68
expeditionary warfare 8, 20, 45, 146, 178, 184–186, 188, 204, 206, 208, 214; Croatia's development of 69; defence transformation towards 211; military capacities related to 21; military means to 190–192; priorities regarding 198; Slovakian capacity for 157, 185; Slovenian strategy of 168; and territorial defence 214
external efforts 7, 28–29, 33, 39, 187, 189, 191, 211–212

F-35 combat aircraft 191, 212
Fenko, Ana Bojinović 169
Ferdinand I of the Habsburg dynasty 61
fighter combat aircraft, self-defence by 55
First World War (WWI) 51; Bulgarian defeat in 51; defeat of Austria-Hungary in 61, 71, 138
Fiszer, Michał 158
force multiplier 17
Fox, Annette Baker 28
free development of the society 106
freedom of enterprise 96
French Empire 161
Fukuyama, Francis 23

Gajdoš, Peter 152
geographical approach, to defence planning 188–193, 200–202; alignment strategy 189; in Baltic States 189; conclusions regarding 193; geographically exposed position 213; great power conflicts and 188; military means 190–192; to multiple-courting 189; offensive bandwagoning strategy 189; strategic ends 189; ways 192–193
geographical characteristics 5, 7, 14, 22–23, 29–30, 182, 188
George, Alexander 11n8
Germany: *Bundeswehr* 179, 185; unification of 11n3
global financial and economic crisis 52, 62
global security system 52
Glváč, Martin 154–155
government in exile 71, 126
grand strategy: definitions of 39, 42; *versus* military strategy 42
Gray, Colin 7, 11n5, 42, 188
Grčić Polić, Jelena 68
Greater Middle East 152
great power club 44

Index

great powers 197; competition among 23; military cooperation (MC) with 32; military strategies of 6; permanent seats in UNSC 25; strategies of 188; tension between 7; *see also* middle powers
Grizold, Anton 169
gross domestic product (GDP) 22, 25, 56
Group of Seven (G7) 43
Group of Twenty (G20) 25, 43

Habsburg Monarchy 93, 115, 126
Hart, Basil Liddell 41
Haushofer, Karl 22
hedging, strategy of 40, 41
Helsinki Head Line Goals (1999) 18, 207
Hendrickson, Ryan 101, 135
historical approach, to defence planning 194–200; alignment strategies 195–197; conclusions regarding 199–200; military means 198–199; strategic ends 197–198; unilateral and/or multilateral approach 199; ways 199
historical experiences 5, 7–10, 14, 20, 22–24, 29–32, 129, 135, 195, 198
Hitler, Adolf 22
Hoffmann, Stanley 30–31
Hojs, Aleš 165
Holy Roman Empire 71, 126, 161
Horthy, Miklós 93
host nation support (HNS) 66
House of Habsburg 71
humanitarian crises 72
Hungarian army: armoured reconnaissance battalion 98; capabilities in domestic matters 101; developments of 179; light infantry battalions 98; Mechanised Infantry Brigade, 5th 98; Mechanised Infantry brigade, 25th 98
Hungary: kingdom of 161; Ottoman conquest of 150; rebellion against the Habsburgs 150; Upper Hungary 150
Hungary, strategy of 93–102; alignment strategy 100; based on multiple-courting 100; contributions to the EUBG 98; difference with Czechia's defence strategies 188; fundamental ends of 95–96; historical background 93–94; for international prestige and influence 96; and invasion of Red Army 94; main military resources 98; means of 97–98; national defence 100; national self-strength 98; on policy towards Russia 101; on position within the EU 101; for preservation of the cultural heritage and identity 96; self-government and autonomy 96; strategic environment 94–95; for sustainable economic growth 96; voluntary reserve system 97, 98; ways of 98–100; Zrínyi 2026 plan for the modernisation of the armed forces 98
Huntington, Samuel 23
hybrid warfare 69, 73, 117
Hyde-Price, Adrian 16, 31, 32

Iancu, Avram-Florian 147
identification-friend-or-foe capability 87
infantry fighting vehicles (IFVs) 75
influence 42, 44, 52, 78, 100; of Croatian democratic political system 64; of cultural and political identity 105; of Czech government 177; differences related to relative power 6; of geography and historical experiences 29–32; goals related to 20, 29; on military reform programs 18; of non-state actors 151; of previous experiences of armed conflicts 30; of Romanian state 146, 184; of Russia in Soviet Eastern and Central Europe 23; of Soviet Union 3; sphere of 4, 23, 72, 83; during systematic wars 31
information technology 105
Innocent IV, Pope 115
institutionalised multilateral cooperation 7
instrumental learning 19, 207
internal efforts 7, 28–29, 39, 187, 211, 214
internal security dynamics 14
International Institute for Strategic Studies (IISS) 10, 93, 98
International Monetary Fund (IMF) 26
international organized crime 151
international peace support operations 54, 58, 66, 68, 178, 180, 206
International Relations (IR) 24–25, 42, 83, 100, 139
International Security Assistance Force (ISAF) 76
international terrorism 52–53, 62, 83, 104, 105, 127, 145, 207, 208; illegal proliferation of arms and technologies 52; non-state actors 16; regional instability caused by 18
interoperability, with the armed forces of friendly states 45
interstate armed conflicts 21, 72, 151, 207, 208; in failed/failing states 72; regional instability caused by 18
invading forces, protection against 23

Iohannis, Klaus 141
Iran, Islamic fundamentalism in 127
Iraq, US-led invasion of 134, 176, 183
Islamic State (IS): US-led Operation Inherent Resolve against 99–100; war against 46n1
isolation, strategy of 40

Jagiellonian dynasty (Hungary) 93
Jakniunaite, Dovile 122
JAS-39 Gripen combat aircraft 75, 98
Jelušič, Ljubica 165, 169
Jervis, Robert 7
Ji Yun Lee 41
Johnston, Alistair Iain 5
Joint Air-to-Surface Standoff Missile system 191

Kahn, Herman 41
Kanis, Pavol 158
Kasekamp, Andres 122
Kjellén, Rudolf 22, 188
knowledge, cumulative development of 11n9
Komorowski, Bronistaw 130
Kopac, Erik 168
Kostov, Ivan 59n2
Kotnik-Dvojmoc, Igor 168
Kreslinš, Kārlis 112

land and air forces, downsizing of 210
Lašas, Ainius 23, 188
Latvian state and society 107
Latvia, strategy of 104–112; air defence and anti-tank capabilities 109; capacity to counter NBC weapons 107; Coast Guard Ship Flotilla 108; crusade against pagans in Northern Europe 104; defence expenditures 190; fundamental ends of 106–107; historical background 104; information technology 105; joint military training with other NATO member states 108; key priority in the defence sector 112; main military resources 109; means of 107–109; membership in NATO and the EU 105; military combat capabilities 107; modern command and control systems 107; multilateral approach 111; multiple-courting hedging strategy 111; on national defence 111; National Guard 108–109; on national security 106; participation in EU-led international missions 110; risk of external military aggression 110; self-defence capabilities 112; strategic environment 104–106; strategic goal after the end of the Cold War 106; on threats caused by foreign intelligence and security services 105; ways of 109–111
Layne, Christopher 29
leash slipping, strategy of 41
Lisbon, Treaty of 21, 88
Lithuania, kingdom of: creation of 115; defence policy 121; economic situation of 119; Grand Duchy of 115; independence from Russia 115; membership of NATO and EU 118; Moscow, Treaty of (1920) 115; warfare with the Christian orders, Poland and the Kiev-Russians 115
Lithuanian armed forces: combat service support units 119; development of 122; infantry battalion battle group 119; "Iron Wolf" mechanized infantry brigade 119; jaeger battalion 119; main military resources of **120**; motorised infantry brigade 118; National Defence Volunteer Forces 118; participation in NATO NRF and the EU Battlegroups 119; reformation of 212; Teutonic Knights 115; transformation of 118
Lithuania, strategy of 115–123; alignment strategy 121; bi- and multilateral defence cooperation 123; concept of territorial defence 121; on consequence of globalisation 116; on deterrence based on defence 120; on fostering citizenship and patriotism 118; fundamental ends of 117–118; historical background 115; on hybrid warfare 117; means of 118–120; on multilateral approach 122; multiple-courting hedging strategy 121; national defence 122; strategic environment 116–117; ways of 120–121; white paper on defence on 118
Livonian Order 82
logistic chains 45, 208
Louis II, King 93

Macierewicz, Antoni 131
Mackinder, Halford 22
Mahečić, Zvonimir 69
main battle tanks (MBTs) 55, 68, 97, 192, 210, 212
Manicom, James 25

Mariana, Terra 104
Marušiak, Juraj 101
Matlary, Janne Haaland 15, 206
means: Bulgarian strategy 54–56; Croatian 64–66; Czech Republic (Czechia) strategy 74–76; element of the military strategy 178–179; Estonian strategy 85–87; to expeditionary warfare 190–192; geographical approach 190–192; historical approach 198–199; Hungary, strategy of 97–98; Latvian strategy 107–109; Lithuanian strategy 118–120; for national defence 179; to national defence 190–192; Polish strategy 129–132; to positional approach 184–186; Romanian strategy 142–144; for self-preservation 39; Slovakia, strategy of 153–155; Slovenia, strategy of 164–166
Mearsheimer, John 43
mechanized brigade 54
Membership Action Plan (MAP) 69, 207
Merger Treaty 3
Meri, Lennart 89
Mi-24 Hind helicopters 65, 97–98
Middle Ages 8, 32
Middle East 52–53, 94, 100, 127, 140, 162, 167
middle powers 183; definition of 25, 34n6; diplomacy 28; major 25; minor 26; Poland 26; *see also* great powers
MiG-21 aircraft 64–66, 143
MiG-29 Fulcrum aircraft 97
military assistance (MA) 31
military blocs: Eastern 14; Western 14
military cooperation (MC), with great powers 32
military force, use of 56
military intelligence battalion 65
military means 184–186
military power resources 26
military strategies 41–46; concept of 39; Croatian 69; defined 42; elements of **46**; ends element of 177–178; *versus* grand strategy 42; of great powers 6; key elements of 6; means element of 178–179; operationalisation of 39; ways element of 179–181; *see also* defence strategies
military strength 10, 39, 97
Mill, John Stuart 11n7
mine countermeasures vessel (MCMV) 119
missile boats 143

missile defence program 79, 134, 143, 176, 189, 202
missile ships, Helsinki-class 69
Mistral surface-to-air missiles 85
Miszczak, Krzysztof 134
Mohács, Battle of (1526) 93
Moldavia 32, 93, 138, 141, 144, 213
Mölder, Holger 90
Molotov–Ribbentrop Pact (1939) 23, 188
monarchy, referendum abolishing 51
Moravian Empire 150
Moscow, Treaty of (1920) 115
Moumoutzis, Kyriakos 19, 207
Müller, Patrick 20, 207
multilateral approaches 45, 58, 68, 77–78, 89, 100, 111, 122, 146, 157, 179, 180, 186
multilateral defence cooperation 17, 123
multilateral expeditionary warfare 17
multilateral risk-sharing 17
Multinational Corps Northeast (MNC-NE) 75
Multinational Land Force (MLF) 66, 167
Multinational Peace Force–South-Eastern Europe (MPF-SEE) 142–143
Multinational Stand-by High Readiness Brigade (SHIRBRIG) 142
multiple-courting, strategy of 41, 175, 183; alignment 180; in Bulgaria 57; in Croatia 67; in Czech Republic 77; in Estonia 88; in Latvia 111; in Lithuania 121; military-oriented approach to 194–195, 199; in Romania 146; security-oriented approaches to 200; in Slovakia 157; in Slovenia 167
Munich Agreement (1938) 23, 188

Napoleonic paradigm 16
national characteristics, variables in 22–34; geographical characteristics 22; influence of geography and historical experiences 29–32; position in the international system 24–29; relative power 22, 24–29; strategic exposure 33–34
national defence 178, 184–186, 188; collective defence 45; development of 44; in Latvia 111; military means to 190–192; priorities regarding 198; unilateral capabilities 45
national identity 139, 177
nationally controlled resources 7
national security 7, 210; in Bulgaria 57; in Croatia 69; cyber aspects of 63; in

Estonia 82, 87; goals related to 20; in Hungary 96; in Latvia 104–106, 110; in Lithuania 116–117; national security strategy (NSS) 51; in Poland 127; post-national security paradigm 4, 15–21; in Slovenia 163
national self-strength 98
national strategic cultures, analysis of 23
national territorial defence, notion of 15, 206
nation-state model of defence 15, 206
NATO Response Force (NRF) 64, 66, 74, 76, 110
natural resources, depletion of 95
Nazi-Germany 8, 51, 115, 138, 150, 194, 214; annexation of Sudetenland 71; eastward expansion 188; invasion of Yugoslavia by 61; *Reichskommissariat Ostland* 82
Nemeth, Bence 102
new powers, rise of 22
niche diplomacy 29
non-aligned states 15
Nordic-Baltic military cooperation 112
Nordic countries, defence strategies of 31
Norkus, Renatas 122
North Atlantic Treaty 75, 88
North Atlantic Treaty Organization (NATO) 3, 14, 99; anti-ballistic defence 154; area of responsibility 57; Baltic Air Policing mission fighters 86; Bulgaria's accession to 53, 58; collective defence system 77, 110, 180, 186; Comprehensive Operations Planning Directive (COPD) 207; Cooperative Cyber Defence Centre of Excellence in Estonia 86; creation of 14; criteria for membership 18; Croatia's accession to 67; defence planning 21; development of defence capabilities 74; engagement in stabilisation missions outside Europe 132; Estonia's accession to 87; expansion in Central Europe 152; funding for developing the Slovenian military infrastructure 177; Germany's membership in 197; ISAF in Afghanistan 177; Joint Response Force 143; KFOR in Kosovo 177; Latvia's membership in 105; leadership of 67; Lithuanian membership of 118; Membership Action Plan (MAP) 18–19, 69, 207; membership of 9–10, 207; military exercises 21; obligations of 106; Operational Concept and Force Structure 146; Operation Resolute Support 99; operations against proliferation of WMD 57; Operation Unified Protector 135; opportunities for 'instrumental learning' 207; Partnership for Peace (PfP) program 18–19, 207; Polish membership of 127; Rapid Deployable Corps based in Italy (NRDC-It) 165; Response Force 88; Romania's membership of 140, 146; SFOR (Stabilisation Force) in Croatia 99; Slovenia's membership of 163, 169; Smart Defence 77; strategy of collective defence in Europe 46n1; *Study on NATO Enlargement* (1995) report 18; Washington summit (1999) 19
nuclear, biological and chemical weapons (NBC) protection 54, 107

Oder–Neisse line 126
offensive realists 42
offensive warfare 8, 31, 32, 195, 197
off-shore balancing, strategy of 29
Operation Active Endeavour 143
Operation Atalanta 143
Operation Enduring Freedom 76
Operation Inherent Resolve 99
Operation Iraqi Freedom 110
Operation Mare Nostrum 167
Operation Resolute Support 99
Operation Unified Protector 135, 143
optional wars 15, 206
organised crime 52–53
organised international crime 16, 127
Organization for Economic Co-operation and Development (OECD) 99
Organization for Security and Co-operation in Europe (OSCE) 19, 52, 74, 99, 157, 167
Ottoman Empire 51, 93, 138, 150
Otzulis, Valdis 112, 122
Ozoliņa, Žaneta 112, 122

Paris, Treaty of (1947) 3, 94
Partnership for Peace (PfP) program 18–19, 207
Paszewski, Tomasz 135
Patriot missile defence system 191, 212
peace support operations (PSOs) 4, 17, 185, 206
Péczeli, Anna 5
Perešin, Anita 69
Permanent Structured Cooperation (PESCO) 56

224 *Index*

Piłsudski, Józef 126
Planning and Review Process (PARP) 19
Poland 26, 30, 183, 198, 202; access to greater military resources 202; annexation of the Vilnius region 115; annexation of the Zaolzie region 71; balance between the uni- and multilateral approaches 181; civil resistance during 1980–1989 126; contributions to international military operations 179; contributions to US-led coalitions 189; cyber warfare capacity 131; Duchy of 126; geographical position of 127; industrial defence capacities 131; kingdom of 115; memberships in NATO and the EU 127; as middle power 187; missile defence agreement with US 135; modernisation of the air defence system 191; National Security Bureau (NSB) 127; occupation by Nazi-Germany and the USSR during WWII 126; offensive bandwagoning strategy 189; Patriot missile defence system 191; position during NATO's Operation Unified Protector over Libya 135; protection of its national heritage and identity 129; as sovereign and democratic state in Central Europe 129; strategic culture 135; war with Czechoslovakia 126
Poland armed forces: development of 130; joint anti-terrorist operations 130; main military resources of **132**; missile defence system 131; modernisation of the air defence system 131; tactical mobility of 131; technical modernisation of 131
Poland, strategy of 126–135; balance of power alignment strategy 135; bandwagoning with the US 135; in defence against external crisis 128; fundamental ends of 128–129; historical background 126; means of 129–132; for national defence 134; objectives of defence policy 128; strategic environment 127–128; ways of 132–133
policy learning, definition of 19
Polish-Lithuanian Commonwealth 115
Polish-Lithuanian Union 115
political extremism 116
political leadership, of small states 7
political pluralism, implementation of 96
political unilateralism 180, 187
Posen, Barry 42

positional approach, to defence planning 200; alignment strategies 183–184; conclusions regarding 187–188; differences in relative power 182–188; military means 184–186; strategic ends 184; ways 186–187
post-communist states 5
post-national security paradigm 4, 209–210
Potsdam Conference (1945) 126
power maximisation, notion of 44
power of states 29
problem solving 5
Procházka, Josef 78–79
Provincial Reconstruction Team (PRT) 76

RBS-70 mobile air defence weapon system 108
realism: classical 24; defensive 42; structural 24
Red Army 82, 94
Reeves, Jeffrey 25
regional balancing, strategy of 40
regional conflicts 53, 62, 68, 76, 128, 139, 151
Regional Security Complex Theory (RSCT) 21
Reiter, Dan 8, 194
relative power, notion of 22
religious-extremist-driven terrorism 139
Riga, Treaty of (1920) 104
Rikveilis, Airis 111
rising power, threat perceptions of 41
Roland air defence missile system 166
Romania: Berlin, Treaty of (1878) 138; contribution to regional stability 142; economic security and foreign investments 147; geographical position 139; integration as a NATO and EU member 140, 146; military security of 141; national identity 139; participation in German invasion of the USSR 138; partnership with the US 147; Russian occupation 138; status as a security provider 142; transformation of the armed forces towards expeditionary capabilities 178; Transylvania and Wallachia principalities 138
Romanian armed forces: air policing mission in the Baltic States 143; American AEGIS Ashore system 143; contributions to the EU-led operation Althea in Bosnia-Herzegovina 143; deployment to Afghanistan 143; downsizing of 178; main military

Index 225

resources of **144**; missions of 142; Operation Active Endeavour 143; operational structure of 142, 144; Operation Atalanta 143; Operation Unified Protector 143; training mission to Mali 143; transformation in accordance with NATO and EU standards 142
Romanian-Hungarian Joint Peacekeeping Battalion 143
Romania, strategy of 138–147; adoption of multiple-courting 146; defense and security policy 147; economical-social shortcomings 139; for economic and social development 140; fundamental ends of 140–142; historical background 138; means of 142–144; strategic environment 138–140; ways of 144–145
Rome, Treaty of (1957) 3
Rončević, Berislav 62
Rostoks, Toms 112
Rothstein, Robert 28
Rublovskis, Raimonds 112
Russia: annexation of the Crimea 4, 78, 117, 140, 177, 212; efforts to strengthening its great power status 140; EU's policy in relation to 41; Hungarian policy towards 101; military intervention in Georgia 4, 7, 21, 191, 208, 212, 213; use of military force for political purposes 21; war against Ukraine 4, 23, 122, 157, 177, 187, 190, 198, 212–213

Šabič, Zlatko 169
Saeima (Latvian Parliament) 105–106, 110
Saint-Germain-en-Laye, Treaty of (1919) 71
Schelling, Thomas 41
Scott, David 123
search and rescue (SAR) capability 85
Second World War (WWII) 4, 51, 175, **196–197**
security dynamics 21, 210; in Europe in the post–Cold War era 15; expeditionary warfare 45; internal 14; new European 14–22, 206–211; new post-national 15–21; return of old European 21–22, 188
security intelligence system 61
security policies, of European countries 16
security strategy: of Croatia 63, 69; of Hungary 96
security threats, evaluation of 75
self-defence capabilities 112, 122

self-help strategies, development of 42
self-help system 39
self-identification, idea of 25
self-preservation, means for 39
Simeon II, Tsar 51
Simon, Jeffrey 19, 207
Slovakian armed forces: defence transformation process 186; expeditionary warfare 185; transformation of 185
Slovakian General Staff (GS) 151
Slovakia, strategy of 27, 150–158, 198; alignment strategy 157; capacity for expeditionary warfare 157; COVID-19 pandemic 152; fundamental ends of 152–153; historical background 150; main military resources and **155**; means of 153–155; multilateral approach 157; multiple-courting hedging strategy 157; National Council (NC) 150; security and defence policy 153; strategic environment 150–152; ways of 156–157
Slovenian armed forces: air defence and communications 164; contribution to the joint military capabilities of the EU 164; crisis response operations 166; enlargement and modernization of the Cerklje ob Krki airfield 166; high-readiness battalion battle group 165; main military resources of **166**; mechanized battalion battle group 166; medium infantry battalion group 165; motorised infantry company 164; Operation Mare Nostrum 167; participation to NATO-led operations 164; professionalization of 169; wartime strength of 164
Slovenian minorities, rights and development of 163
Slovenia, strategy of 161–169, 198; expeditionary warfare 168; fundamental ends of 163–164; historical background 161; key objectives of defence policy 163; means of 164–166; membership of NATO and EU 163, 169; multilateral approach 168; multiple-courting hedging strategy 167; passive bandwagoning 168; position and reputation in the international community 163; strategic environment 161–163; War of Independence 161; ways of 166–167
small states 6–7, 23, 25, 90, 182–184, 187; defence strategies 27–28; European 17;

goals related to influence and status 29; historical experiences relating to WWI and WWII 31; institutionalised multilateral cooperation 7; multilateral strategies for collective territorial defence 214; multiple-courting strategy 200; national characteristics 22; operationalisation of influence 43; participation in international PSOs 208; political leadership of 7; strategic exposure 33; strategic priorities related to national defence among 190; unilateral approaches 210
Smith, Rubert 16
Snyder, Jack 11n5
Sobotka, Bohuslav 73
social development 63, 73, 94, 140, 163
socialization, notion of 20
social learning 19–20, 207
social stratification 105
social terrorism 116
South Eastern Europe Brigade (SEEBRIG) 143
sovereignty and integrity, of the state 177
Soviet military equipment, dependence of 21
Soviet Union 8, 197; Estonians war of independence against 82; German invasion of 138; implosion of 15; invasion of Bulgaria 51; occupation of Czechoslovakia 71; power competition with US 14; *see also* Russia
special forces battalion 54, 65
special operations forces squadron 119, 191
state-building processes 5, 8, 24, 30
states' competition, for status and recognition 43
state's political independence and territorial integrity, protection of 43
status 8; definition of 44; great power 140; of Latvian language 105; of middle power 25; of national economy 104; peace support operations (PSOs) 17; positional 43; of Romania as a security provider 142; of Russia as global power 83; states' competition for 43; survival 177, 184, 189
stealth capacities, for warships and airplanes 45
Stephen II, King 61
Stephen I, King 93
Stockholm International Peace Research Institute (SIPRI) 26
strategic culture, definition of 5, 11n5
strategic ends 184

strategic exposure: and future challenges facing the new European allies 213–215; notion of 33–34; potential aggregated perception of **33**
strategic resources, supply of 207
Strategic Studies 41–42
structural realism, notion of 24
structured focused comparison (SFC) 6, 9
submarine warfare 65
surface-to-air missiles 75
survival 7, 24, 43–44, 57, 175, 177–178, 184, 187, 189; categories of 39; of Czech state 78; defensive realists 42; of Estonia 89; goals related to 20; of Hungary 96, 100; of Latvian state 111; of Lithuania 121; military strategy for 68, 89; against military threats 78; in nuclear age 41; of Poland 134; of Romania 146; of Slovakia 157; of Slovenian state 168; supremacy of 58; threats of 28
sustainable economic development 53
Svinarov, Nikolay 52–53
Swedish Empire 126

Takacs, David 89
Tartu, Treaty of (1920) 82
Tashev, Blagovest 58
Taylor, Maxwell 39, 42
technologically advanced military systems 45
technological revolution 95
territorial defence 86, 111, 121
territorial integrity 184
territorial security 194
Teutonic Knights 115
totalitarian ideologies 16
transatlantic security 99
trans-border organized crime 62
transport aircrafts 108
transport systems, for deployment of military units 45
transport technology 22
Transylvania, Principality of 93
Treaty on Conventional Armed Forces in Europe 75
Treaty on European Union 76
Trianon, Treaty of (1920) 93–94
Truman doctrine 23

unilateral approach 134, 179–181, 186–187, 199, 208, 210
UN Interim Force in Lebanon (UNIFIL) 167

Index 227

unipolar power, notion of 40
United Nations (UN) 17, 52
United Nations Peacekeeping Force in Cyprus (UNFICYP) 99
United States (US): anti-terrorist operation 76; ballistic missile defence programme 176; containment policy against world communism 188; invasion of Iraq 176; military presence in Latvia 110; missile defence program 79; Operation Enduring Freedom 76; Operation Inherent Resolve 99; Operation Iraqi Freedom 110; power competition with USSR 14; relation with Croatia 69; unipolar power 40
universal human values 53
UN Security Council (UNSC) 25, 43, 134
Upper Hungary, Principality of 150
Urbanovská, Jana 158
use of force, policy of 42
USSR *see* Russia; Soviet Union
Ustaše organisation 61

Vaicekauskaitė, Živilė Marija 123
Vanaga, Nora 112
Varg, Gergely 101
Velvet Revolution (1989) 71
Venice, Republic of 161
Versailles, Treaty of 126
Vienna Document 75
Visegrád Cooperation 100, 157
Visegrád countries 98
Visegrád Group 5
Vojtek, Peter 155
voluntary reserve system 97–98
von Clausewitz, Carl 41

Walt, Stephen 40–41
Waltz, Kenneth 39
Warsaw Pact (WP) 5, 101; creation of 14; defence planning of 21, 207; dissolution of 15, 59; establishment of 51; founding members of 51; resolution of 11n3
War Ship Flotilla 108
Washington Treaty (1946) 3, 55, 74, 130
Watkins, Amadeo 69
ways: of Bulgaria 56–57; of Croatia 66–67; of Czech Republic (Czechia) 76–77; element of the military strategy 179–181; of Estonia 87–88; geographical approach 192–193; historical approach 199; of Hungary 98–100; of Latvia 109–111; of Poland 132–133; positional approach 186–187; Romania's priorities regarding 144–145, 189; of Slovakia 156–157; of Slovenia 166–167; unilateral and/or multilateral approach 186–187, 192–193
weapons of mass destruction (WMD) 18, 53, 62, 72, 94, 105, 116, 127, 139, 151, 207
Western military bloc 14; *see also* North Atlantic Treaty Organization (NATO)
Wirtz, James 41
World Trade Organization (WTO) 89

Yalta Agreement (1945) 23, 188
Yaniszewski, Mark 101, 135
Yugoslavia, Kingdom of 61, 161; Axis invasions of 93; wars of independence (1991–2001) 206
Yugoslav regency 61

Zajac, Justyna 5
Załęski, Krzysztof 134
Zartaloudis, Sotirios 19, 207
Zeitschrift für Geopolitik 22
Zord, Gábor 101
Zrínyi 2026 plan, for the modernisation of the armed forces 98
ZU-23-2 anti-aircraft cannons 85